Statistical Tables

and Formulas

WILEY PUBLICATIONS
IN STATISTICS

Walter A. Shewhart, Editor

Statistical Tables
and
Formulas

A. HALD

Professor of Statistics
University of Copenhagen

New York · John Wiley & Sons, Inc.
London · Chapman & Hall, Limited

PRINTED IN THE UNITED STATES OF AMERICA

Preface

THE PRESENT COLLECTION OF TABLES AND FORMULAS IS A REVISED VERSION of the appendix to my Danish textbook on statistics, published in 1948, which has now been published in English.

A survey of the most important formulas given in the textbook is presented, and tables for use in connection with these formulas are given. Together with the formulas, references are given to the textbook where proofs and examples may be found.

The principles used for tabulating the distributions are those of R. A. Fisher. The tables contain 21 fractiles for the t-, χ^2-, and W-distributions, and 17 for the v^2-distribution, representing an extension of previously published tables. Linear interpolation both with respect to probability level and number of degrees of freedom will be sufficient for most applications in practice. Also distribution and power curves may easily be drawn from the tables, in particular when using probability or logarithmic probability paper.

I am indebted to Professor R. A. Fisher, Dr. F. Yates, and Messrs. Oliver and Boyd, Ltd., Edinburgh, for kind permission to reproduce some tables from their book *Statistical Tables*, and to Professor E. S. Pearson and the proprietors of *Biometrika* for kind permission to reproduce some tables published in *Biometrika*. Adequate references are given in connection with each table.

A. HALD

UNIVERSITY OF COPENHAGEN
October, 1951

5

CONTENTS

Formulas

Tables

REFERENCES ARE TO A. HALD: STATISTICAL THEORY WITH ENGINEERING APPLICATIONS.

JOHN WILEY & SONS, NEW YORK, 1952.

8

FORMULAS

Notation and Fundamental Formulas.

In one-dimensional distributions the stochastic variable is usually denoted by x and the observed values by $x_1, x_2, \ldots, x_n$. The *number of observations in a sample* is denoted by n and the *number of samples* by k.

The symbol $P\{a < x \leq b\}$ denotes *the probability that the inequality $a < x \leq b$ is satisfied*.

The *distribution function $p\{x\}$* and the *cumulative distribution function $P\{x\}$* satisfy the following relationships:

a. For continuous distributions

$$P\{x\} = \int_{-\infty}^{x} p\{x\}dx \quad \text{or} \quad \frac{dP\{x\}}{dx} = p\{x\}, \quad \S \ 5.2, \text{ p. 91.}$$

b. For discontinuous distributions

$$P\{x_\nu\} = \sum_{i=-\infty}^{\nu} p\{x_i\} \quad \text{or} \quad p\{x_\nu\} = P\{x_\nu\} - P\{x_{\nu-1}\}, \quad \S \ 5.3, \text{ p. 95.}$$

The *fractile x_P* is defined by the relationship $P\{x \leq x_P\} = P$, § 5.6, p. 102.

THE PROBABILITY P FOUND IN THE TABLES OF FRACTILES IS ALWAYS THE PROBABILITY RECKONED FROM THE LOWER LIMIT OF THE DISTRIBUTION.

In formulas involving limits of significance or confidence limits which contain two fractiles, x_{P_1} and x_{P_2}, P_1 is always less than P_2.

In most cases *Greek letters* are used for *parameters or population values* and corresponding *Roman letters* for their *estimates or sample values*.

The mean of x, the *population mean*, is defined as

$$\xi = M\{x\} = \int_{-\infty}^{\infty} xp\{x\}dx \quad \text{or} \quad \sum_{i=-\infty}^{\infty} x_i p\{x_i\}, \quad \S \ 5.7, \text{ p. 103.}$$

The mean of a transformed variable, $y = \varphi(x)$, is not equal to the transformed value of the mean of the original variable unless the transformation is linear, i. e., $M\{\varphi(x)\} \neq \varphi(M\{x\})$ unless $\varphi(x) = a + \beta x$. However, if y is approximately linear over practically the whole range of variation of x, then $M\{\varphi(x)\} \simeq \varphi(\xi)$, § 5.8, p. 105.

The variance of x, the *population variance*, is defined as

$$\sigma^2 = V\{x\} = M\{(x-\xi)^2\}, \quad \S \ 5.9, \text{ p. 106.}$$

The positive square root of the variance is called the *standard deviation*. For a linear function of x we have $V\{a+\beta x\} = \beta^2 V\{x\}$. If $y = \varphi(x)$ is approximately linear over practically the whole range of variation of x it follows that $V\{\varphi(x)\} \simeq (\varphi'(\xi))^2 V\{x\}$, § 5.9, p. 106.

The *coefficient of variation* is defined as $\gamma = \sigma/\xi$, § 5.9, p. 106.

Measuring a stochastic variable from its mean, using the standard deviation as unit, leads to a *standardized variable* $u = (x-\xi)/\sigma$, having *zero mean and unit variance*, § 5.9, p. 106.

In two-dimensional distributions the stochastic variables are usually denoted by (x, y) and the observed values by $(x_1, y_1), (x_2, y_2), \ldots, (x_n, y_n)$. (In Chapter 19 on the two-dimensional normal distribution the variables have been denoted by (x_1, x_2)).

The *distribution function* $p\{x, y\}$ and the *cumulative distribution function* $P\{x, y\}$ satisfy the following relationship:

a. For continuous distributions

$$P\{x, y\} = \int_{-\infty}^{x} \int_{-\infty}^{y} p\{x, y\} dx dy , \quad \S\ 5.12,\ \text{p. 111.}$$

b. For discontinuous distributions

$$P\{x_i, y_j\} = \sum_{\nu=-\infty}^{i} \sum_{\mu=-\infty}^{j} p\{x_\nu, y_\mu\} , \quad \S\ 5.13,\ \text{p. 113.}$$

The *marginal distribution* of x is

$$p\{x\} = \int_{-\infty}^{\infty} p\{x, y\} dy \quad \text{or} \quad p\{x_i\} = \sum_{j=-\infty}^{\infty} p\{x_i, y_j\} .$$

The *conditional distributions* $p\{x|y\}$ and $p\{y|x\}$ satisfy the relationship

$$p\{x, y\} = p\{x\} \cdot p\{y|x\} = p\{y\} \cdot p\{x|y\} .$$

The *conditional mean and variance* of y for given x are defined as the mean and variance of y in the conditional distribution of y for given x, i. e.,

$$m\{y|x\} = \int_{-\infty}^{\infty} y p\{y|x\} dy$$

and

$$v\{y|x\} = \int_{-\infty}^{\infty} (y-m\{y|x\})^2 p\{y|x\} dy , \quad \S\ 5.15,\ \text{p. 114.}$$

The conditional mean of y for given x is called the *regression* of y on x. If

$$m\{y|x\} = \alpha+\beta(x-\bar{x}) = \alpha_0+\beta x ,$$

the regression is linear with the *regression coefficient* β.

The covariance of (x, y) is defined as

$$v\{x, y\} = m\{(x-\xi)(y-\eta)\} , \quad \text{where} \quad \xi = m\{x\} \quad \text{and} \quad \eta = m\{y\} , \quad \S\ 5.16,\ \text{p. 115,}$$

from which we get the *correlation coefficient* as

$$\varrho = \varrho(x, y) = v\{x, y\}/\sqrt{v\{x\}v\{y\}} , \quad \S\ 5.16,\ \text{p. 115.}$$

If $z = a_0+a_1x+a_2y$, then

$$m\{z\} = a_0+a_1 m\{x\}+a_2 m\{y\}$$

and

$$\mathcal{V}\{z\} = a_1^2\mathcal{V}\{x\}+a_2^2\mathcal{V}\{y\}+2a_1a_2\mathcal{V}\{x, y\}\ , \quad \S\ 5.17,\ \text{p. 117.}$$

If $z = \varphi(x, y)$ is approximately linear over practically the whole range of variation of (x, y), then

$$\mathcal{M}\{z\} \simeq \varphi(\xi, \eta),\ \text{where}\ \xi = \mathcal{M}\{x\}\ \text{and}\ \eta = \mathcal{M}\{y\},$$

and

$$\mathcal{V}\{z\} \simeq (\varphi_x'(\xi, \eta))^2\,\mathcal{V}\{x\}+(\varphi_y'(\xi, \eta))^2\,\mathcal{V}\{y\}+2\varphi_x'(\xi, \eta)\varphi_y'(\xi, \eta)\,\mathcal{V}\{x, y\}, \quad \S\ 5.17,\ \text{p. 117.}$$

Estimates or sample values corresponding to the above-mentioned parameters or population values are defined as follows:

A (relative) *frequency* is denoted by $h = x/n$ and a cumulative frequency by H.

The *sample mean* is defined as

$$\bar{x} = \frac{1}{n}\sum_{i=1}^{n}x_i\ , \quad \S\ 4.3,\ \text{p. 67.}$$

Introducing a suitably chosen computing origin a and unit β the original observations are linearly transformed to $w_i = (x_i-a)/\beta$ and $\bar{w}$ is computed, whence $\bar{x} = a+\beta\bar{w}$, $\S$ 4.3, p. 67.

The *sample variance* is defined as

$$s_x^2 = \frac{1}{n-1}\sum_{i=1}^{n}(x_i-\bar{x})^2 = \frac{1}{n-1}\left(\sum_{i=1}^{n}x_i^2 - \frac{1}{n}\left(\sum_{i=1}^{n}x_i\right)^2\right), \quad \S\ 4.4,\ \text{p. 72.}$$

The positive square root of the variance is called the *standard deviation*. For a linear transformation of the observations $x_i = a+\beta w_i$ we have $s_x^2 = \beta^2 s_w^2$, $\S$ 4.4, p. 72.

The *covariance* is defined as

$$s_{xy} = \frac{1}{n-1}\sum_{i=1}^{n}(x_i-\bar{x})(y_i-\bar{y}) = \frac{1}{n-1}\left(\sum_{i=1}^{n}x_iy_i - \frac{1}{n}\sum_{i=1}^{n}x_i\sum_{i=1}^{n}y_i\right), \quad \S\ 4.8,\ \text{p. 85.}$$

Linear transformations $x_i = a+\beta v_i$ and $y_i = \gamma+\varkappa w_i$ give $s_{xy} = \beta\varkappa s_{vw}$.

As a check on the computation of $\bar{x}$, $\bar{y}$, s_x^2, s_y^2, and s_{xy} we may introduce an auxiliary variable $z_i = a_0+a_1x_i+a_2y_i$, $i = 1, 2, \ldots, n$, with suitably chosen coefficients and compute $\bar{z}$ and s_z^2, which should satisfy the relations $\bar{z} = a_0+a_1\bar{x}+a_2\bar{y}$ and $s_z^2 = a_1^2s_x^2+a_2^2s_y^2+2a_1a_2s_{xy}$, $\S$ 4.8, p. 85.

The *correlation coefficient* is found as $r_{xy} = s_{xy}/s_xs_y$, $\S$ 4.8, p. 85.

The estimated linear regression of y on x is $Y = \bar{y}+b(x-\bar{x})$, where the *sample regression coefficient* b is computed as

$$b = \frac{\displaystyle\sum_{i=1}^{n}(x_i-\bar{x})(y_i-\bar{y})}{\displaystyle\sum_{i=1}^{n}(x_i-\bar{x})^2}\ , \quad \S\ 18.3,\ \text{p. 528.}$$

The *range*, $w = w_n$, denotes the difference between the largest and the smallest observation in a sample, $\S$ 12.1, p. 319.

The standard deviation of a function of the observations is often called the *standard error*. For example, $\sigma_{\bar{x}} = \sigma/\sqrt{n}$ denotes the theoretical standard error of the sample mean $\bar{x}$, and $s_{\bar{x}} = s/\sqrt{n}$ denotes the corresponding empirical standard error.

Further, the following notation is used for computational purposes, see $\S$ 4.4, p. 72, and $\S$ 4.8, p. 85: $S = Sum$, $SS = Sum\ of\ Squares$, $SP = Sum\ of\ Products$, $SSD = Sum\ of\ Squares\ of\ Devia$-

tions, and $SPD = Sum$ of $Products$ of $Deviations$, where the deviations always are meant as deviations from the corresponding least square estimates. For example we have

$$SSD_x = \sum (x_i - \bar{x})^2 \qquad SPD_{xy} = \sum (x_i - \bar{x})(y_i - \bar{y}).$$

WHEN NOTHING IS SAID TO THE CONTRARY THE OBSERVATIONS ARE ASSUMED TO BE NORMALLY DISTRIBUTED AND STOCHASTICALLY INDEPENDENT.

Formulas for Tests of Significance.

The formulas for tests of significance given in the following usually involve three terms: (1) the *critical region*, i. e., the region which leads to rejection of the test hypothesis; (2) the *power function* denoted by π, i. e., the probability of rejecting the test hypothesis as a function of the unknown parameter; and (3) the *number of observations* necessary to discriminate between two given values of the parameter with the power required. The probability of rejecting the test hypothesis when it is true is called the *level of significance* and is usually denoted by a. To discriminate between two values, ξ_0 and ξ_1 say, so that the probability of rejecting the hypothesis $\xi = \xi_0$ when it is true is a, and similarly the probability of rejecting $\xi = \xi_1$ when this hypothesis is true is β, a certain number of observations is required. This number may be determined from the equations $\pi(\xi_0) = a$ and $\pi(\xi_1) = 1 - \beta$ where $\pi(\xi)$ denotes the power function for testing the hypothesis $\xi = \xi_0$ against the alternative $\xi = \xi_1$.

Only formulas for one-sided tests at the a level of significance are given. The corresponding formulas for two-sided tests based on "equal tail areas" may be derived by substituting $a/2$ for a, which gives the critical region and the power function for the one tail, and adding a similar term for the other tail.

Some Fundamental Distribution Functions.

The binomial distribution (Chapter 21):

$$p\{x\} = \binom{n}{x} \theta^x (1-\theta)^{n-x}, \quad x = 0, 1, \ldots, n .$$

$$m\{x\} = n\theta \quad \text{and} \quad v\{x\} = n\theta(1-\theta) .$$

The Poisson distribution (Chapter 22):

$$p\{x\} = e^{-\xi} \frac{\xi^x}{x!}, \quad x = 0, 1, \ldots .$$

$$m\{x\} = v\{x\} = \xi .$$

The hypergeometric distribution (§§ 2.4, 17.2, and 21.8):

$$p\{x\} = \frac{\binom{M}{x} \binom{N-M}{n-x}}{\binom{N}{n}}, \quad x = 0, 1, \ldots, m ,$$

where m equals the smaller of the two integers n and M.

$$\mathcal{M}\{x\} = n\theta \quad \text{and} \quad \mathcal{V}\{x\} = n\theta(1-\theta)\left(1 - \frac{n-1}{N-1}\right) \quad \text{where} \quad \theta = \frac{M}{N}.$$

The multinomial distribution (Chapter 23):

$$p\{x_1, x_2, \ldots, x_k\} = \frac{n!}{x_1! \, x_2! \ldots x_k!} \, \theta_1^{x_1}\theta_2^{x_2}\ldots\theta_k^{x_k}, \; \Sigma\theta_i = 1, \; 0 \leq x_i \leq n, \text{ and } \Sigma x_i = n \, .$$

The normal distribution (Chapter 6):

$$p\{x\} = \frac{1}{\sqrt{2\pi}\,\sigma} \, e^{-\frac{(x-\xi)^2}{2\sigma^2}}, \quad -\infty < x < \infty \, .$$

$$\mathcal{M}\{x\} = \xi \quad \text{and} \quad \mathcal{V}\{x\} = \sigma^2 \, .$$

The two-dimensional normal distribution (Chapter 19):

$$p\{x_1, x_2\} = \frac{1}{2\pi\,\sigma_1\sigma_2\sqrt{1-\varrho^2}} \, e^{-\frac{1}{2(1-\varrho^2)}\left[\left(\frac{x_1-\xi_1}{\sigma_1}\right)^2 - 2\varrho\,\frac{x_1-\xi_1}{\sigma_1}\frac{x_2-\xi_2}{\sigma_2} + \left(\frac{x_2-\xi_2}{\sigma_2}\right)^2\right]}, \quad -\infty < x_i < \infty, \; i = 1, 2 \, .$$

$$\mathcal{M}\{x_i\} = \xi_i, \; \mathcal{V}\{x_i\} = \sigma_i^2, \text{ and } \mathcal{V}\{x_1, x_2\} = \varrho\sigma_1\sigma_2 \, .$$

$$\mathcal{M}\{x_2|x_1\} = \xi_2 + \varrho\,\frac{\sigma_2}{\sigma_1}(x_1-\xi_1) \quad \text{and} \quad \mathcal{V}\{x_2|x_1\} = \sigma_2^2(1-\varrho^2) \, .$$

The χ^2-distribution (Chapter 10):

$$p\{\chi^2\} = \frac{1}{2^{\frac{f}{2}}\,\Gamma\left(\frac{f}{2}\right)} \, (\chi^2)^{\frac{f}{2}-1} \, e^{-\frac{\chi^2}{2}}, \quad 0 \leq \chi^2 < \infty \, .$$

The v^2-distribution (Chapter 14):

$$p\{v^2\} = \frac{\Gamma\left(\frac{f_1+f_2}{2}\right)}{\Gamma\left(\frac{f_1}{2}\right)\Gamma\left(\frac{f_2}{2}\right)} \, f_1^{\frac{f_1}{2}} f_2^{\frac{f_2}{2}} \, \frac{(v^2)^{\frac{f_1}{2}-1}}{(f_2+f_1 v^2)^{\frac{f_1+f_2}{2}}}, \quad 0 \leq v^2 < \infty \, .$$

The t-distribution (Chapter 15):

$$p\{t\} = \frac{1}{\sqrt{\pi f}} \, \frac{\Gamma\left(\frac{f+1}{2}\right)}{\Gamma\left(\frac{f}{2}\right)} \left(1 + \frac{t^2}{f}\right)^{-\frac{f+1}{2}}, \quad -\infty < t < \infty \, .$$

Planning of Sampling Investigations.

Random Sampling From a Finite Population, § 17.2, p. 488.

For the sample mean we have $\mathcal{M}\{\bar{x}\} = \xi$ and

$$\mathcal{V}\{\bar{x}\} = \frac{\sigma^2}{n}\left(1 - \frac{n-1}{N-1}\right) \simeq \frac{\sigma^2}{n}(1-a), \quad \text{where} \quad a = \frac{n}{N} \, .$$

13

14

Stratified Random Sampling, § 17.3, p. 491.

Dividing a population into k sub-populations or strata in the proportions $w_1 : w_2 : \ldots : w_k$, $\Sigma w_i = 1$, we have

$$p\{x\} = \sum_{i=1}^{k} w_i p_i\{x\}, \quad m\{x\} = \xi = \sum_{i=1}^{k} w_i \xi_i,$$

and

$$\mathcal{V}\{x\} = \sigma^2 = \sum_{i=1}^{k} w_i \sigma_i^2 + \sum_{i=1}^{k} w_i (\xi_i - \xi)^2,$$

where $p_i\{x\}$, ξ_i, and σ_i^2 denote the distribution function, the mean, and the variance of the ith sub-population.

Drawing a random sample of size n_i from stratum number i the weighted mean $\bar{x} = \Sigma w_i \bar{x}_i$ will be an unbiased estimate of ξ with variance

$$\mathcal{V}\{\bar{x}\} = \sum_{i=1}^{k} w_i^2 \frac{\sigma_i^2}{n_i} (1 - a_i), \quad a_i = \frac{n_i}{N_i}.$$

By proportional sampling, i. e., the sampling fractions are equal for all strata, we obtain

$$\mathcal{V}\{\bar{x}\} = \frac{\Sigma w_i \sigma_i^2}{n} (1 - a), \quad a = \Sigma n_i / \Sigma N_i.$$

For a given total sample size, $n = \Sigma n_i$, the minimum variance of $\bar{x}$ is obtained by choosing

$$n_i = \frac{w_i \sigma_i}{\Sigma w_i \sigma_i} n, \quad i = 1, 2, \ldots, k,$$

leading to

$$\mathcal{V}\{\bar{x}\} = \frac{(\Sigma w_i \sigma_i)^2}{n} - \frac{\Sigma w_i \sigma_i^2}{N} = \frac{(\Sigma w_i \sigma_i)^2}{n} \left(1 - a \frac{\Sigma w_i \sigma_i^2}{(\Sigma w_i \sigma_i)^2} \right).$$

Taking costs into consideration the minimum value of $\mathcal{V}\{\bar{x}\}$ under the assumption of given cost, $c = c_0 + \Sigma c_i n_i$, will be obtained for

$$n_i = \frac{w_i \sigma_i / \sqrt{c_i}}{\Sigma w_i \sigma_i / \sqrt{c_i}} n.$$

Two-Stage Sampling, § 17.4, p. 496.

From a population consisting of N_1 primary units each containing N_2 secondary units we select a random sample of n_1 primary units and from each of these a random sample of n_2 secondary units. Characterizing the variation between the means of primary units by $\sigma_1^2 = \Sigma (\xi_i - \xi)^2 / N_1$ and the average variation within primary units by $\sigma_2^2 = \Sigma \Sigma (X_{iv} - \xi_i)^2 / N_1 N_2$, the variance of the sample mean becomes

$$\mathcal{V}\{\bar{x}\} = \frac{\sigma_1^2}{n_1} (1 - a_1) + \frac{\sigma_2^2}{n_1 n_2} (1 - a_2), \quad a_1 = \frac{n_1}{N_1} \quad \text{and} \quad a_2 = \frac{n_2}{N_2}.$$

If the cost is c_1 per primary unit and c_2 per secondary unit the total cost becomes $c = c_1 n_1 + c_2 n_1 n_2$. Minimizing $\mathcal{V}\{\bar{x}\}$ for given total cost leads to

$$n_2 = \frac{\sigma_2}{\sigma_1} \sqrt{\frac{c_1}{c_2}} \bigg/ \sqrt{1 - \frac{\sigma_2^2}{N_2 \sigma_1^2}}.$$

Plans for testing hypotheses regarding specified population parameters may be derived from the corresponding power functions which will be given in the following.

Some Applications of the Tables of the Normal Distribution, Tables I, II and III.

APPLICATIONS TO THE NORMAL DISTRIBUTION.

The *standardized normal distribution function* given in Table I is defined as

$$\varphi(u) = \frac{1}{\sqrt{2\pi}} e^{-\frac{u^2}{2}}, \quad \varphi(-u) = \varphi(u), \quad \S\ 6.3,\ \text{p. 121.}$$

The standardized variable u has zero mean and unit variance.

The *standardized normal cumulative distribution function* given in Table II is defined as

$$\Phi(u) = \frac{1}{\sqrt{2\pi}} \int_{-\infty}^{u} e^{-\frac{x^2}{2}} dx, \quad \Phi(-u) = 1 - \Phi(u), \quad \S\ 6.3,\ \text{p. 121.}$$

The *fractiles* u_P of the standardized normal variable are defined by the equation

$$\frac{1}{\sqrt{2\pi}} \int_{-\infty}^{u_P} e^{-\frac{x^2}{2}} dx = P, \quad u_{1-P} = -u_P, \quad \S\ 6.6,\ \text{p. 127.}$$

Table III contains some values of $u_P + 5$.

Let x be normally distributed with mean ξ and variance σ^2, i. e., $x = \xi + \sigma u$. The distribution of x may be tabulated by means of the u-distribution as follows:

The distribution function:

$$p\{x\} = \frac{1}{\sqrt{2\pi}\,\sigma} e^{-\frac{(x-\xi)^2}{2\sigma^2}} = \frac{1}{\sigma} \varphi(u), \quad u = \frac{x-\xi}{\sigma}, \quad \S\ 6.4,\ \text{p. 124.}$$

The cumulative distribution function:

$$P\{x\} = \frac{1}{\sqrt{2\pi}\,\sigma} \int_{-\infty}^{x} e^{-\frac{(x-\xi)^2}{2\sigma^2}} dx = \Phi(u), \quad u = \frac{x-\xi}{\sigma}, \quad \S\ 6.5,\ \text{p. 125.}$$

The fractiles:

$$x_P = \xi + \sigma u_P, \quad \S\ 6.6,\ \text{p. 127.}$$

The construction of *fractile diagrams* to test the normality of empirical distributions, § 6.7, p. 130:

a. Ungrouped observations: abscissa $x_{(i)}$, the ith smallest observation, ordinate $u_{\frac{i-1/2}{n}} + 5$.

b. Grouped observations: abscissa $t_j + \frac{\Delta t_j}{2}$, the jth class limit, ordinate $u_{H_j} + 5$.

The "expected" number of observations in the jth class interval when a normal distribution has been fitted to the data with mean $\bar{x}$ and standard deviation s is

$$n\left[\Phi\left(\frac{t_j+\dfrac{\varDelta t_j}{2}-\bar{x}}{s}\right)-\Phi\left(\frac{t_j-\dfrac{\varDelta t_j}{2}-\bar{x}}{s}\right)\right],\quad \S\ 6.8,\ \text{p. 140.}$$

The Distribution of the Sample Mean.

The mean $\bar{x}$ is normally distributed with parameters $(\xi,\ \sigma^2/n)$. It follows that

$$P\left\{\xi+u_{P_1}\frac{\sigma}{\sqrt{n}}<\bar{x}<\xi+u_{P_2}\frac{\sigma}{\sqrt{n}}\right\}=P_2-P_1,\quad \S\ 9.3,\ \text{p. 221,}$$

and

$$P\left\{\bar{x}-u_{P_2}\frac{\sigma}{\sqrt{n}}<\xi<\bar{x}-u_{P_1}\frac{\sigma}{\sqrt{n}}\right\}=P_2-P_1,\quad \S\ 9.5,\ \text{p. 239.}$$

Testing the Hypothesis $\xi \leq \xi_0$ Against the Alternative $\xi > \xi_0$, $\S$ 9.4, p. 224.

At the α level of significance the critical region is $\dfrac{\bar{x}-\xi_0}{\sigma/\sqrt{n}}>u_{1-\alpha}$ with the power function

$$\pi(\xi)=\Phi(u_\alpha+\lambda),\quad \text{where}\quad \lambda=\frac{\xi-\xi_0}{\sigma/\sqrt{n}}.$$

The number of observations necessary to discriminate between ξ_0 and ξ_1, $\xi_0<\xi_1$, so that $\pi(\xi_0)=\alpha$ and $\pi(\xi_1)=1-\beta$ is given by

$$\sqrt{n}=(u_{1-\beta}+u_{1-\alpha})\frac{\sigma}{\xi_1-\xi_0}.$$

Transformation of Skew Distributions, Chapter 7.

If $g(x)$ is normally distributed with mean $g(\xi)$ and variance σ^2, then

$$p\{x\}=\frac{g'(x)}{\sigma}\varphi(u),\quad \text{and}\quad P\{x\}=\Phi(u),\quad \text{where}\quad u=\frac{g(x)-g(\xi)}{\sigma}.$$

In particular: If $\log x$ is normally distributed with mean $\log\xi$ and variance σ^2, then

$$p\{x\}=\frac{0\cdot4343}{\sigma x}\varphi(u),\quad \text{and}\quad P\{x\}=\Phi(u),\quad \text{where}\quad u=\frac{\log x-\log\xi}{\sigma},\quad \S\ 7.2,\ \text{p. 160,}$$

which is called the *logarithmic normal* distribution.

Some Approximation Formulas based on the Normal Distribution.

For moderately large values of n (the number of observations) or f (the number of degrees of freedom) the exact sampling distributions of some estimates may be approximated to by normal distributions with sufficient accuracy for most practical purposes.

The Distribution of the Standard Deviation.

The standard deviation s is approximately normally distributed for $f>30$ with mean $\sigma\sqrt{1-\dfrac{1}{2f}}$

and variance $\dfrac{\sigma^2}{2f}$, § 11.7, p. 299. It follows that

$$u \simeq \frac{s}{\sigma}\sqrt{2f} - \sqrt{2f-1}, \quad \text{for} \quad f > 30,$$

from which limits for s for given σ and f may be found as

$$P\left\{\frac{\sigma}{\sqrt{2f}}\left(\sqrt{2f-1}+u_{P_1}\right) < s < \frac{\sigma}{\sqrt{2f}}\left(\sqrt{2f-1}+u_{P_2}\right)\right\} \simeq P_2 - P_1.$$

Confidence limits for σ are given by

$$P\left\{\frac{s\sqrt{2f}}{\sqrt{2f-1}+u_{P_2}} < \sigma < \frac{s\sqrt{2f}}{\sqrt{2f-1}+u_{P_1}}\right\} \simeq P_2 - P_1.$$

The exact distribution of s^2 and limits for s^2 and σ^2 are given on p. 23.

The Distribution of the Coefficient of Variation.

The coefficient of variation $c = s/\bar{x}$ is approximately normally distributed with mean $\gamma = \sigma/\xi$ and variance $\gamma^2(1+2\gamma^2)/2f$ for large values of f and small values of γ, § 11.8, p. 301.

The Distribution of the Transformed Correlation Coefficient.

The transformed correlation coefficient z is approximately normally distributed with mean ζ and variance $1/(n-3)$, i. e., $u \simeq (z-\zeta)\sqrt{n-3}$, where

$$z = 1\cdot 1513 \log_{10}\frac{1+r}{1-r} \quad \text{and} \quad \mathcal{M}\{z\} = \zeta \simeq 1\cdot 1513 \log_{10}\frac{1+\varrho}{1-\varrho} + \frac{\varrho}{2(n-1)}, \quad \text{§ 19.12, p. 608.}$$

The equality of two population correlation coefficients may be tested by means of the u-test since

$$u \simeq \frac{z_1 - z_2}{\sqrt{\dfrac{1}{n_1-3}+\dfrac{1}{n_2-3}}}, \quad \text{§ 19.12, p. 608.}$$

The Binomial Distribution.

Two sets of approximate formulas are given, the one based on the binomial variable itself, the other on the arc sine transformation.

a. The absolute frequency x is approximately normally distributed with mean $n\theta$ and variance $n\theta(1-\theta)$, from which it follows that the (relative) frequency $h = x/n$ is approximately normally distributed with mean θ and variance $\theta(1-\theta)/n$. This approximation holds good for $n\theta(1-\theta) > 9$. Taking the correction for continuity into account we have

$$P\{x\} \simeq \Phi(u) \quad \text{for} \quad u = \frac{x+\frac{1}{2}-n\theta}{\sqrt{n\theta(1-\theta)}}$$

or

$$P\{h\} \simeq \Phi(u) \quad \text{for} \quad u = \frac{h + \dfrac{1}{2n} - \theta}{\sqrt{\dfrac{\theta(1-\theta)}{n}}}, \quad \S\ 21.5,\ \text{p. 676.}$$

Further

$$h_P + \frac{1}{2n} \simeq \theta + u_P \sqrt{\frac{\theta(1-\theta)}{n}}, \quad \S\ 21.5,\ \text{p. 676.}$$

b. The transformed relative frequency $y = 2 \arcsin \sqrt{h}$, see Table XII, is approximately normally distributed with mean $2 \arcsin \sqrt{\theta}$ and variance $1/n$ for $n\theta(1-\theta) > 9$. It follows that

$$P\{h\} \simeq \Phi(u) \quad \text{for} \quad u = \left(2 \arcsin \sqrt{h + \frac{1}{2n}} - 2 \arcsin \sqrt{\theta}\right)\sqrt{n}$$

and

$$2 \arcsin \sqrt{h_P + \frac{1}{2n}} \simeq 2 \arcsin \sqrt{\theta} + \frac{u_P}{\sqrt{n}}, \quad \S\ 21.6,\ \text{p. 685.}$$

The mean of the two results (a and b) will usually be a good approximation to the exact value.

Correspondingly, the upper and lower confidence limits for θ may be computed from the two formulas

a.

$$\left.\begin{array}{l}\bar{\theta} \\ \underline{\theta}\end{array}\right\} \simeq \frac{1}{n + u_P^2}\left[x \pm \frac{1}{2} + \frac{u_P^2}{2} - u_P\sqrt{\frac{(x \pm \frac{1}{2})(n - x \mp \frac{1}{2})}{n} + \frac{u_P^2}{4}}\right]$$

and

b.

$$\left.\begin{array}{l}2 \arcsin \sqrt{\bar{\theta}} \\ 2 \arcsin \sqrt{\underline{\theta}}\end{array}\right\} \simeq 2 \arcsin \sqrt{\frac{x \pm \frac{1}{2}}{n} - \frac{u_P}{\sqrt{n}}},$$

where the upper sign corresponds to $\bar{\theta}$, the lower to $\underline{\theta}$, and P is replaced by P_1 and P_2, respectively, $\S\ 21.11$, p. 697. The exact confidence limits are given on p. 29, and the 95% and 99% limits have been tabulated in Table XI.

Testing the hypothesis $\theta \leq \theta_0$ against the alternative $\theta > \theta_0$, $\S\ 21.10$, p. 694.

a. At the a level of significance the critical region is

$$\frac{h - \dfrac{1}{2n} - \theta_0}{\sigma_0} > u_{1-a} \quad \text{for} \quad \sigma_0 = \sqrt{\frac{\theta_0(1-\theta_0)}{n}}$$

The power function becomes

$$\pi(\theta) \simeq \Phi\left((u_a + \lambda)\frac{\sigma_0}{\sigma}\right) \quad \text{for} \quad \lambda = \frac{\theta - \theta_0}{\sigma_0} \quad \text{and} \quad \sigma = \sqrt{\frac{\theta(1-\theta)}{n}}.$$

The number of observations necessary to discriminate between θ_0 and θ_1, $\theta_0 < \theta_1$, so that $\pi(\theta_0) = a$ and $\pi(\theta_1) = 1 - \beta$ is given by

$$\sqrt{n} \simeq \frac{u_{1-\beta}\sqrt{\theta_1(1-\theta_1)}+u_{1-\alpha}\sqrt{\theta_0(1-\theta_0)}}{\theta_1-\theta_0}.$$

b. Using the arc sine transformation the critical region becomes

$$\left(2\arcsin\sqrt{h-\frac{1}{2n}}-2\arcsin\sqrt{\theta_0}\right)\sqrt{n} > u_{1-\alpha}.$$

The power function is

$$\pi(\theta) \simeq \Phi(u_\alpha+\lambda), \quad \text{where} \quad \lambda = \left(2\arcsin\sqrt{\theta}-2\arcsin\sqrt{\theta_0}\right)\sqrt{n},$$

and the number of observations necessary to discriminate between θ_0 and θ_1 so that $\pi(\theta_0) = \alpha$ and $\pi(\theta_1) = 1-\beta$ may be found from the equation

$$\sqrt{n} \simeq \frac{u_{1-\beta}+u_{1-\alpha}}{2\arcsin\sqrt{\theta_1}-2\arcsin\sqrt{\theta_0}}.$$

Testing the hypothesis $\theta_1 = \theta_2$ against the alternative $\theta_1 > \theta_2$ on the basis of two observed frequencies $h_1 = \dfrac{x_1}{n_1} \approx \theta_1$ and $h_2 = \dfrac{x_2}{n_2} \approx \theta_2$, § 21.13, p. 705.

a. At the α level of significance the critical region is

$$\frac{\left(h_1-\dfrac{1}{2n_1}\right)-\left(h_2+\dfrac{1}{2n_2}\right)}{\sqrt{\dot{h}(1-h)\left(\dfrac{1}{n_1}+\dfrac{1}{n_2}\right)}} > u_{1-\alpha}, \quad \text{where} \quad h = \frac{x_1+x_2}{n_1+n_2}.$$

b. Using the arc sine transformation the critical region becomes

$$\frac{2\arcsin\sqrt{h_1-\dfrac{1}{2n_1}}-2\arcsin\sqrt{h_2+\dfrac{1}{2n_2}}}{\sqrt{\dfrac{1}{n_1}+\dfrac{1}{n_2}}} > u_{1-\alpha}$$

with the power function

$$\pi(\delta) = \Phi(u_\alpha+\lambda), \quad \text{where} \quad \lambda = \frac{\delta}{\sigma} = \frac{2\arcsin\sqrt{\theta_1}-2\arcsin\sqrt{\theta_2}}{\sqrt{\dfrac{1}{n_1}+\dfrac{1}{n_2}}}.$$

The number of observations necessary to discriminate between θ_1 and θ_2 so that $\pi(0) = \alpha$ and $\pi(\delta) = 1-\beta$ may be found from the equation

$$\sqrt{\frac{1}{n_1}+\frac{1}{n_2}} = \frac{2\arcsin\sqrt{\theta_1}-2\arcsin\sqrt{\theta_2}}{u_{1-\beta}+u_{1-\alpha}}.$$

For $n_1 = n_2 = m$ we have

$$\sqrt{\frac{m}{2}} = \frac{u_{1-\beta}+u_{1-\alpha}}{2\arcsin\sqrt{\theta_1}-2\arcsin\sqrt{\theta_2}}.$$

The Poisson Distribution.

Two sets of approximate formulas are given, the one based on the Poisson variable itself, the other on the square root transformation.

a. The Poisson variable x is approximately normally distributed with mean ξ and variance ξ for $\xi > 9$. It follows that

$$P\{x\} \simeq \Phi(u), \quad \text{where} \quad u = \frac{x + \frac{1}{2} - \xi}{\sqrt{\xi}}, \quad \text{for} \quad \xi > 9, \quad \S\ 22.2,\ \text{p.}\ 717.$$

b. The transformed variable $y = 2\sqrt{x}$ is approximately normally distributed with mean $2\sqrt{\xi}$ and unit variance, so that

$$P\{x\} \simeq \Phi(u), \quad \text{where} \quad u = 2\left(\sqrt{x + \frac{1}{2}} - \sqrt{\xi}\right), \quad \text{for} \quad \xi > 9, \quad \S\ 22.2,\ \text{p.}\ 717.$$

The mean of the corresponding approximations to the fractile is

$$x_P \simeq \xi - \tfrac{1}{2} + u_P \sqrt{\xi} + \tfrac{1}{8} u_P^2.$$

The critical region for testing the hypothesis $\xi \leq \xi_0$ against the alternative $\xi > \xi_0$ is

$$x > \xi_0 + \tfrac{1}{2} + u_{1-\alpha} \sqrt{\xi_0} + \tfrac{1}{8} u_{1-\alpha}^2.$$

Correspondingly, the upper and lower *confidence limits* for ξ may be computed from the formulas

a.
$$\left.\begin{array}{c}\bar{\xi} \\ \underline{\xi}\end{array}\right\} \simeq x \pm \frac{1}{2} + \frac{u_P^2}{2} - u_P \sqrt{x \pm \frac{1}{2} + \frac{u_P^2}{4}}, \quad \S\ 22.5,\ \text{p.}\ 722.$$

b.
$$\left.\begin{array}{c}\bar{\xi} \\ \underline{\xi}\end{array}\right\} \simeq x \pm \frac{1}{2} + \frac{u_P^2}{4} - u_P \sqrt{x \pm \frac{1}{2}}, \quad \S\ 22.5,\ \text{p.}\ 722.$$

Exact values based on the χ^2-distribution will be given on p. 24.

Testing the hypothesis $\xi_1 = \xi_2$ against the alternative $\xi_1 > \xi_2$ on the basis of two observed Poisson-distributed frequencies $x_1 \approx \xi_1$ and $x_2 \approx \xi_2$, $\S\ 22.7$, p. 725.

The critical region is

$$\frac{x_1 - x_2 - 1}{\sqrt{x_1 + x_2}} > u_{1-\alpha}.$$

Tests of Randomness.
The distribution of the number of runs.

$R =$ the total number of *runs above and below the sample median*. The hypothesis of statistical control or randomness may be tested against the alternative hypothesis of shifts in the level of the distribution by means of R, a small number of runs being significant. The distribution of R is approximately normal for $n > 20$ with mean $(n+2)/2$ and variance $(n-1)/4$, $\S\ 13.3$, p. 342.

$R =$ the total number of *runs up and down*. The hypothesis of statistical control may be tested against the alternative hypothesis of a gradual change (trend or cycle) in the level of the distribution by means of R, a small number of runs being significant. The distribution of R is approximately normal for $n > 20$ with mean $(2n-1)/3$ and variance $(16n-29)/90$, $\S\ 13.4$, p. 353.

A test based on *the mean square successive difference*.

Defining

$$q^2 = \frac{1}{2(n-1)} \sum_{i=1}^{n-1} (x_{i+1} - x_i)^2$$

we may use the ratio $r = q^2/s^2$ to test the hypothesis of statistical control against the alternative hypothesis of a gradual change in the population mean, small values of r being significant. The distribution of r is approximately normal for $n > 20$ with mean 1 and variance $1/(n+1)$, § 13.5, p. 357.

Some Tables of the Normal Distribution.

More extensive tables of the normal distribution than the one given here may be found in the following publications:

N. R. Jørgensen: *Frequensflader og Korrelation*, A. Busck, Copenhagen, 1916.

K. Pearson: *Tables for Statisticians and Biometricians*, I, 1924, and II, 1931, Biometrika Office, London.

W. F. Sheppard: *The Probability Integral*, British Association for the Advancement of Science, Mathematical Tables, VII, Cambridge, 1939.

National Bureau of Standards: *Tables of Probability Functions*, I, Mathematical Tables 8, Washington, 1941.

Some Applications of the Table of the *t*-Distribution, Table IV.

The comparison of an observed value of a normally distributed variable and a hypothetical (true) value may be based on the *t*-test, if an s^2-distributed estimate of the variance is given. The quantity distributed as t is $t = (x - \xi)/s_x$, where x and s_x must be stochastically independent, § 15.1, p. 388. The following table gives a number of stochastic variables (x) and corresponding hypothetical values (ξ) and standard errors (s_x) from which the *t*-test may be derived.

The critical region for testing the hypothesis $\xi = \xi_0$ against the alternative $\xi > \xi_0$ at the α level of significance is given by $t > t_{1-\alpha}$. An approximation to the power function for moderately large values of f may be found as

$$\pi(\xi) \simeq \Phi\left(\frac{t_\alpha + \lambda}{\sqrt{1 + \frac{t_{1-\alpha}^2}{2f}}}\right) \quad \text{for} \quad \lambda = \frac{\xi - \xi_0}{\sigma/\sqrt{n}}, \quad \text{§ 15.2, p. 391.}$$

Confidence limits for ξ are given by

$$P\{x - t_{P_2}s_x < \xi < x - t_{P_1}s_x\} = P_2 - P_1.$$

The hypothesis $\varrho = 0$ in a two-dimensional normal distribution may be tested by computing $t = \frac{r}{\sqrt{1-r^2}}\sqrt{f}$, $f = n - 2$, § 19.12, p. 608.

In the comparison of two estimates, $\bar{x}_1$ and $\bar{x}_2$ say, the corresponding variances s_1^2 and s_2^2 have been pooled giving the estimate s^2 as shown in the table below. However, if $\sigma_1^2 \neq \sigma_2^2$, which may be investigated by calculating the variance ratio s_1^2/s_2^2, the *t*-test should be replaced by another test

Stochastic variable	Symbol (x)	Hypothetical value (ξ)
Mean	$\bar{x}$	ξ
Mean difference	$\bar{d} = \bar{x}_1 - \bar{x}_2$	δ
Difference between two means	$\bar{x}_1 - \bar{x}_2$	$\xi_1 - \xi_2$
Mean of dependent variable in regression analysis with one independent variable	$\bar{y}$	α
Regression coefficient in regression analysis with one independent variable	b	β
Value of dependent variable in regression equation with one independent variable	$Y = \bar{y} + b(x - \bar{x})$	$\eta = \alpha + \beta(x - \bar{x})$
Difference between two regression coefficients in regression equations with one independent variable	$b_1 - b_2$	$\beta_1 - \beta_2$
Regression coefficient (m independent variables)	b_i	β_i

based on the quantity

$$t \simeq \frac{\bar{x}_1 - \bar{x}_2 - \delta}{\sqrt{\dfrac{s_1^2}{n_1} + \dfrac{s_2^2}{n_2}}}, \quad \text{where} \quad \frac{1}{f} = \frac{c^2}{f_1} + \frac{(1-c)^2}{f_2} \quad \text{and} \quad c = \frac{\dfrac{s_1^2}{n_1}}{\dfrac{s_1^2}{n_1} + \dfrac{s_2^2}{n_2}}, \quad \S\ 15.4, \text{p. } 394,$$

which is approximately distributed as t with f degrees of freedom, f being defined by the above equation.

Some Applications of the Tables of the χ^2-Distribution, Tables V and VI.

The variable χ^2 is defined as the sum of squares of f stochastically independent u-values, $\chi^2 = \sum_{i=1}^{f} u_i^2$, f being called the number of degrees of freedom. The mean of χ^2 is f, § 10.2, p. 254.

The Distribution of the Variance.

Application of the method of least squares in the analysis of variance and regression leads to a *sum of squares of residuals* which divided by the number of degrees of freedom gives the sample variance s^2. The number of degrees of freedom is equal to the number of residuals minus the number

Standard error (s_x)	Definition of s^2	Degrees of freedom (f)	References
$\dfrac{s}{\sqrt{n}}$	$\dfrac{1}{n-1}\sum(x_i-\bar{x})^2$	$n-1$	§ 15.2, p. 391
$\dfrac{s}{\sqrt{n}}$	$\dfrac{1}{n-1}\sum(d_i-\bar{d})^2$	$n-1$	§ 15.5, p. 401
$s\sqrt{\dfrac{1}{n_1}+\dfrac{1}{n_2}}$	$\dfrac{1}{n_1+n_2-2}\Big[\sum(x_{1i}-\bar{x}_1)^2+\sum(x_{2i}-\bar{x}_2)^2\Big]$	n_1+n_2-2	§ 15.4, p. 394
$\dfrac{s}{\sqrt{n}}$	$\dfrac{1}{n-2}\sum(y_i-Y_i)^2$	$n-2$	§ 18.4, p. 540
$\dfrac{s}{\sqrt{\sum(x_i-\bar{x})^2}}$	»	»	»
$s\sqrt{\dfrac{1}{n}+\dfrac{(x-\bar{x})^2}{\sum(x_i-\bar{x})^2}}$	»	»	»
$s\sqrt{\dfrac{1}{\sum(x_{1i}-\bar{x}_1)^2}+\dfrac{1}{\sum(x_{2i}-\bar{x}_2)^2}}$	$\dfrac{1}{n_1+n_2-4}\Big[\sum(y_{1i}-Y_{1i})^2+\sum(y_{2i}-Y_{2i})^2\Big]$	n_1+n_2-4	§ 18.8, p. 571
$s\sqrt{c_{ii}}$	$\dfrac{1}{n-m-1}\sum(y_i-Y_i)^2$	$n-m-1$	§ 20.3, p. 638

of linear constraints between them. The distribution of the variance can be derived from the χ^2-distribution by the transformation $s^2 = \sigma^2\chi^2/f$, § 11.1, p. 276.

For given σ^2 and f, limits for s^2 may be found as

$$P\left\{\sigma^2\frac{\chi^2_{P_1}}{f} < s^2 < \sigma^2\frac{\chi^2_{P_2}}{f}\right\} = P_2-P_1, \quad \text{§ 11.2, p. 278.}$$

Confidence limits for σ^2 are given by

$$P\left\{s^2\frac{f}{\chi^2_{P_2}} < \sigma^2 < s^2\frac{f}{\chi^2_{P_1}}\right\} = P_2-P_1, \quad \text{§ 11.4, p. 286.}$$

Testing the hypothesis $\sigma^2 \leq \sigma_0^2$ against the alternative $\sigma^2 > \sigma_0^2$, § 11.3, p. 280.

At the a level of significance the critical region is $s^2 > \sigma_0^2\chi^2_{1-a}/f$ with the power function

$$\pi(\sigma^2) = P\left\{\chi^2 > \frac{\sigma_0^2}{\sigma^2}\chi^2_{1-a}\right\}.$$

The number of degrees of freedom necessary to discriminate between σ_0^2 and σ_1^2, $\sigma_0^2 < \sigma_1^2$, so that $\pi(\sigma_0^2) = a$ and $\pi(\sigma_1^2) = 1-\beta$ may be found by solving the equation

$$\frac{\chi^2_{1-a}}{\chi^2_\beta} = \frac{\sigma_1^2}{\sigma_0^2}.$$

For moderately large values of f an approximate solution is given by

$$f \simeq \tfrac{1}{2} + \tfrac{1}{2}\left(\frac{u_{1-\beta}\lambda + u_{1-\alpha}}{\lambda - 1}\right)^2, \quad \text{where} \quad \lambda = \frac{\sigma_1}{\sigma_0}.$$

The Addition Theorem for the s^2-Distribution, § 11.5, p. 287.

From k sample variances, $s_1^2, s_2^2, \ldots, s_k^2$, with the same theoretical value σ^2 we may calculate a pooled estimate s^2 of σ^2 as the weighted average of the k variances using the numbers of degrees of freedom as weights, i. e., $s^2 = \sum f_i s_i^2 / \sum f_i$. In particular, from k duplicate measurements with the same variance σ^2 we may compute a pooled estimate s^2 as the sum of squares of the k differences divided by $2k$, the number of degrees of freedom for this estimate being k.

Testing the Equality of k Population Variances, BARTLETT's *Test*, § 11.6, p. 290.

From the k sample variances we compute the pooled estimate s^2 with $f = \sum f_i$ degrees of freedom and

$$\chi^2 \simeq \frac{2 \cdot 3026}{c}\left(f \log s^2 - \sum_{i=1}^{k} f_i \log s_i^2\right), \quad \text{where} \quad c = 1 + \frac{1}{3(k-1)}\left(\sum_{i=1}^{k}\frac{1}{f_i} - \frac{1}{f}\right),$$

which quantity is approximately distributed as χ^2 with $k-1$ degrees of freedom, large values of χ^2 being significant.

Testing the Equality of k Probabilities on the Basis of k Observed Binomially Distributed Frequencies, § 21.14, p. 711.

$$h_1 = \frac{x_1}{n_1}, \ldots, h_k = \frac{x_k}{n_k}, \overline{h} = \frac{\Sigma x}{\Sigma n}. \qquad \chi^2 \simeq \frac{1}{\overline{h}(1-\overline{h})}\sum_{i=1}^{k} n_i(h_i - \overline{h})^2, \ f = k-1.$$

The Poisson Distribution.

The cumulative distribution function may be expressed by the χ^2-distribution since

$$P\{x\} = e^{-\xi}\sum_{\nu=0}^{x}\frac{\xi^\nu}{\nu!} = 1 - P\{\chi^2 < 2\xi\} \text{ for } f = 2(x+1), \quad \text{§ 22.1, p. 714.}$$

The *fractiles* may be found as $x_P = \dfrac{f}{2} - 1$, where f is determined from the equation $\chi_{1-P}^2 = 2\xi$, x_P being an integer.

Confidence limits for ξ for an observed x may be found as

$$\overline{\xi}(x) = \tfrac{1}{2}\chi_{1-P_1}^2, \ f = 2(x+1), \quad \text{and} \quad \underline{\xi}(x) = \tfrac{1}{2}\chi_{1-P_2}^2, \ f = 2x, \quad \text{§ 22.5, p. 722.}$$

Comparing k Poisson-distributed observations:

$$\chi^2 \simeq \frac{\displaystyle\sum_{i=1}^{k}(x_i - \overline{x})^2}{\overline{x}}, \ f = k-1, \quad \text{§ 22.8, p. 726.}$$

Single Sampling Inspection Plans Based on the Poisson-Distribution, § 21.12, p. 700.

Rule of action: Accept the lot if a random sample of n items contains c or less defective items;

otherwise reject the lot. It is assumed that the sample contains less than 10% of the items in the lot and the proportion of defective items in the lot is less than 0·1. Denoting the proportion defective in the lot by θ, the power function becomes

$$\pi(\theta) = P\{x \geq c+1\} \simeq P\{\chi^2 < 2n\theta\} \quad \text{for} \quad f = 2(c+1) \, .$$

For a required degree of quality assurance expressed by $\pi(\theta_1) = \alpha$ and $\pi(\theta_2) = 1-\beta$ the acceptance number c may be found as $c = \dfrac{f}{2} - 1$, where f is determined as the number of degrees of freedom satisfying the equation $\chi^2_{1-\beta}/\chi^2_\alpha = \theta_2/\theta_1$. The sample size is found as $n = \chi^2_\alpha/2\theta_1$.

Comparison of an Observed Grouped Distribution and a Theoretical Distribution.

A random sample of n observations is distributed into k classes according to the probabilities $\theta_1, \theta_2, \ldots, \theta_k, \Sigma\theta_i = 1$. The number of observations falling into the ith class is denoted by a_i. A test for the agreement between the observed and "expected" numbers is

$$\chi^2 \simeq \sum_{i=1}^{k} \frac{(a_i - n\theta_i)^2}{n\theta_i}, \quad n\theta_i > 5, \, f = k-1 \, , \quad \S \, 23.1, \text{ p. 739.}$$

If the probabilities are estimated from the data the test becomes

$$\chi^2 \simeq \sum_{i=1}^{k} \frac{(a_i - n\hat{\theta}_i)^2}{n\hat{\theta}_i}, \quad n\hat{\theta}_i > 5, \, f = k-c-1 \, , \quad \S \, 23.2, \text{ p. 742,}$$

c linear relations between the deviations being introduced by the estimation.

A Test for Stochastical Independence by Two-Way Classification, $\S \, 23.3,$ p. 744.

A random sample of n observations is distributed according to two criteria into $m \times k$ classes, the number of observations in the ijth class being a_{ij}. Denoting the corresponding marginal numbers by a_{i0} and a_{0j} the test is based upon the quantity

$$\chi^2 \simeq n \sum_{i=1}^{m} \sum_{j=1}^{k} \frac{\left(a_{ij} - \dfrac{a_{i0}a_{0j}}{n}\right)^2}{a_{i0}a_{0j}}, \quad \text{where} \quad n = \sum\sum a_{ij} \quad \text{and} \quad f = (m-1)(k-1) \, .$$

Testing the Equality of k Population Correlation Coefficients, $\S \, 19.12,$ p. 608.

Transforming the k correlation coefficients, $r_1, \ldots, r_k$, we find

$$z_i = 1·1513 \log_{10} \frac{1+r_i}{1-r_i}, \quad \text{and} \quad \bar{z} = \sum(n_i - 3)z_i / \sum(n_i - 3) \, .$$

The hypothesis is then tested by calculating $\chi^2 \simeq \sum_{i=1}^{k} (n_i - 3)(z_i - \bar{z})^2 \, , \, f = k-1.$

Tolerance Limits, $\S \, 11.10,$ p. 311.

If P denotes the probability that at least the fraction $1 - 2\varepsilon$ of the normal distribution will be included between the limits $\bar{x} \pm sl$, then

$$l \simeq u_{1-\varepsilon} \sqrt{\frac{f}{\chi^2_{1-P}}} \left(1+\frac{1}{2n}\right).$$

The one-sided tolerance limit corresponding to the proportion $1-\theta$ of the population is $\bar{x}+st_P(\theta)$, where

$$t_P(\theta) \simeq \frac{u_{1-\theta}+u_P \sqrt{\frac{1}{n}\left(1-\frac{u_P^2}{2f}\right)+\frac{u_{1-\theta}^2}{2f}}}{1-\frac{u_P^2}{2f}}.$$

Tables of Fractiles of the χ^2-Distribution:

R. A. FISHER and F. YATES: *Statistical Tables*, Oliver and Boyd, London, 1943. Fractiles to three decimal places for $f = 1(1)30$.

C. M. THOMPSON: *Table of Percentage Points of the χ^2-Distribution*, Biometrika, 32, 1941, 187–191. Fractiles to six significant figures for $f = 1(1)30(10)100$.

A. HALD and S. A. SINKBÆK: *A Table of Percentage Points of the χ^2-Distribution*, Skandinavisk Aktuarietidskrift, 1950, 168–175. Fractiles to three decimal places for $f = 1(1)100$.

Some Applications of the Table of the v^2-Distribution, Table VII.

The variable v^2 is defined as the ratio between two sample variances having the same true value

$$v^2 = \frac{s_1^2}{s_2^2} = \frac{\chi_1^2/f_1}{\chi_2^2/f_2}, \quad \S\ 14.1,\ \text{p. } 374.$$

The table contains fractiles only for $P \geq 0.5$ since fractiles for $P < 0.5$ may be calculated from the formula $v_P^2(f_1, f_2) = 1/v_{1-P}^2(f_2, f_1)$.

In most cases linear interpolation will suffice; for more accurate results use $1/f_1$ and $1/f_2$ as arguments.

Testing the Hypothesis $\sigma_1^2 \leq \sigma_2^2$ Against the Alternative $\sigma_1^2 > \sigma_2^2$, § 14.2, p. 379.

At the α level of significance the critical region is $\dfrac{s_1^2}{s_2^2} > v_{1-\alpha}^2(f_1, f_2)$ with the power function

$$\pi(\lambda^2) = P\left\{v^2(f_1, f_2) > \frac{1}{\lambda^2} v_{1-\alpha}^2(f_1, f_2)\right\}, \quad \text{where} \quad \lambda^2 = \frac{\sigma_1^2}{\sigma_2^2}.$$

The number of degrees of freedom necessary to discriminate between given values of σ_1^2 and σ_2^2 so that $\pi(\lambda^2) = 1-\beta$ may be found by solving the equation

$$\lambda^2 = v_{1-\alpha}^2(f_1, f_2) v_{1-\beta}^2(f_2, f_1).$$

Confidence limits for the ratio σ_1^2/σ_2^2 are given by

$$P\left\{\frac{s_1^2}{s_2^2} \frac{1}{v_{P_2}^2(f_1, f_2)} < \frac{\sigma_1^2}{\sigma_2^2} < \frac{s_1^2}{s_2^2} \frac{1}{v_{P_1}^2(f_1, f_2)}\right\} = P_2-P_1, \quad \S\ 14.3,\ \text{p. } 381.$$

A Test for the Equality of the Means of k Normal Populations with the Same Variance, § 16.4, p. 423.

The following table summarizes the analysis of variance for k sets of observations.

Variation	SSD	f	s^2	$m\{s^2\}$	Test
Between sets	$\sum_{i=1}^{k} n_i(\bar{x}_i - \bar{x})^2$	$k-1$	s_2^2	$\sigma^2 + \dfrac{\Sigma n_i(\xi_i - \bar{\xi})^2}{k-1}$	$v^2 = \dfrac{s_2^2}{s_1^2}$
Within sets	$\sum_{i=1}^{k} \sum_{\nu=1}^{n_i} (x_{i\nu} - \bar{x}_i)^2$	$\sum_{i=1}^{k} n_i - k$	s_1^2	σ^2	
Total	$\sum_{i=1}^{k} \sum_{\nu=1}^{n_i} (x_{i\nu} - \bar{x})^2$	$\sum_{i=1}^{k} n_i - 1$			

The hypothesis is tested by the variance ratio $v^2 = s_2^2/s_1^2$, large values of v^2 being significant. An approximation to the power function may be found as

$$\pi(\lambda^2) \simeq P\left\{ v^2(f_2', f_1) > \frac{1}{\lambda^2} v_{1-\alpha}^2(f_2, f_1) \right\},$$

where

$$\lambda^2 = \frac{m\{s_2^2\}}{m\{s_1^2\}} \quad \text{and} \quad f_2' = f_2 \frac{\lambda^4}{2\lambda^2 - 1}.$$

Analysis of the Variance Into Two Components Representing Different Sources of Random Variation, § 16.5, p. 437.

The analysis of variance is carried out as shown in the table above with the only modification that

$$m\{s_2^2\} = \sigma^2 + \bar{n}\omega^2, \quad \text{where} \quad \bar{n} = \frac{1}{k-1}\left(\Sigma n_i - \frac{\Sigma n_i^2}{\Sigma n_i} \right),$$

ω^2 denoting the variance of the normally distributed population means.

For all $n_i = n$ we have $m\{s_2^2\} = \sigma^2 + n\omega^2$. In this case the hypothesis $\omega^2 = 0$ may be tested from the variance ratio $v^2 = s_2^2/s_1^2$, the power function being

$$\pi(\lambda^2) = P\left\{ v^2(f_2, f_1) > \frac{1}{\lambda^2} v_{1-\alpha}^2(f_2, f_1) \right\}, \quad \lambda^2 = 1 + \frac{n\omega^2}{\sigma^2}, \quad f_1 = k(n-1), \quad \text{and} \quad f_2 = k-1,$$

and correspondingly

$$\lambda^2 = 1 + \frac{n\omega^2}{\sigma^2} = v_{1-\alpha}^2(f_2, f_1) v_{1-\beta}^2(f_1, f_2).$$

A generalization of the above analysis of variance to multi-stage grouping may be found in § 16.6, p. 447.

The Analysis of Variance by Two-Way Classification.

The arithmetic of the analysis has been summarized in the table below.

By means of variance ratios different hypotheses may be tested. Whether to use s_1^2 or s_2^2 as denominator depends on the interpretation of the data, see § 16.7, p. 456.

Variation	SSD	f	s^2
Between columns	$nk \sum\limits_{j=1}^{m} (\bar{x}_{.j} - \bar{x}_{..})^2$	$m-1$	s_4^2
Between rows	$nm \sum\limits_{i=1}^{k} (\bar{x}_{i.} - \bar{x}_{..})^2$	$k-1$	s_3^2
Interaction	$n \sum\limits_{i=1}^{k} \sum\limits_{j=1}^{m} (\bar{x}_{ij} - \bar{x}_{i.} - \bar{x}_{.j} + \bar{x}_{..})^2$	$(m-1)(k-1)$	s_2^2
Within sets	$\sum\limits_{i=1}^{k} \sum\limits_{j=1}^{m} \sum\limits_{\nu=1}^{n} (x_{ij\nu} - \bar{x}_{ij})^2$	$km\,(n-1)$	s_1^2
Total	$\sum\limits_{i=1}^{k} \sum\limits_{j=1}^{m} \sum\limits_{\nu=1}^{n} (x_{ij\nu} - \bar{x}_{..})^2$	$kmn-1$	

A generalization to m-way classification has been given in § 16.8, p. 480.

A Test of Linearity for a Regression Equation with One Independent Variable, § 18.3, p. 528.

$$s_1^2 = \frac{1}{\Sigma n_i + k} \sum_{i=1}^{k} \sum_{\nu=1}^{n_i} (y_{i\nu} - \bar{y}_i)^2, \quad s_2^2 \doteq \frac{1}{k-2} \sum_{i=1}^{k} n_i (\bar{y}_i - Y_i)^2, \quad v^2 = \frac{s_2^2}{s_1^2}.$$

Testing the Equality of k Regression Coefficients in Regression Equations with One Independent Variable, § 18.9, p. 579.

$s_1^2 =$ the sum of squares of residuals from the k regression lines divided by the number of degrees of freedom, $f_1 = \sum n_i - 2k$.

$$s_2^2 = \frac{1}{k-1} \sum_{i=1}^{k} (b_i - \bar{b})^2 SSD_{x_i}, \quad v^2 = \frac{s_2^2}{s_1^2}.$$

Further tests regarding the comparison of k regression lines are given in § 18.9, p. 579.

Testing a Hypothesis Regarding the Regression Coefficients in a Linear Regression Equation with m Independent Variables, § 20.2 and § 20.3, p. 627 and p. 638.

$$s_1^2 = \frac{1}{n-m-1} \sum_{\nu=1}^{n} (y_\nu - Y_\nu)^2, \quad s_2^2 = \frac{1}{m} \left[\sum_{i=1}^{m} (b_i - \beta_i)^2 SSD_{x_i} + 2 \sum_{i=1}^{m-1} \sum_{j=i+1}^{m} (b_i - \beta_i)(b_j - \beta_j) SPD_{x_i x_j} \right], \quad v^2 = \frac{s_2^2}{s_1^2}$$

Testing a Hypothesis Regarding the Population Means in a Two-Dimensional Normal Population, § 19.11, p. 606.

$$T^2 = \frac{1}{1-r^2} \left[\left(\frac{\bar{x}_1 - \xi_1}{s_{\bar{x}_1}} \right)^2 - 2r \frac{\bar{x}_1 - \xi_1}{s_{\bar{x}_1}} \frac{\bar{x}_2 - \xi_2}{s_{\bar{x}_2}} + \left(\frac{\bar{x}_2 - \xi_2}{s_{\bar{x}_2}} \right)^2 \right] = 2v^2(2, \ n-2) \frac{n-1}{n-2}.$$

Testing the Equality of Two Sets of Population Means in Two-Dimensional Normal Populations with the Same Variances and Covariances, § 19.16, p. 616.

$$d_i = \bar{x}_i - \bar{y}_i, \quad s_{d_i}^2 = s_i^2 \left(\frac{1}{n_1} + \frac{1}{n_2} \right), \quad i = 1, 2,$$

and

$$T^2 = \frac{1}{1-r^2}\left[\left(\frac{d_1}{s_{d_1}}\right)^2 - 2r\frac{d_1}{s_{d_1}}\frac{d_2}{s_{d_2}} + \left(\frac{d_2}{s_{d_2}}\right)^2\right] = 2v^2(2,\ n_1+n_2-3)\frac{n_1+n_2-2}{n_1+n_2-3}.$$

The Cumulative Binomial Distribution Function may be expressed as

$$P\{x\} = \sum_{v=0}^{x}\binom{n}{v}\theta^v(1-\theta)^{n-v} = 1-P\left\{v^2<\frac{n-x}{x+1}\frac{\theta}{1-\theta}\right\},\ f_1 = 2(x+1)\ \text{ and }\ f_2 = 2(n-x)\,,$$

§ 21.2, p. 672.

Confidence limits for θ:

$$\bar{\theta} = \frac{(x+1)v_{1-P_1}^2}{n-x+(x+1)v_{1-P_1}^2},\ \ \begin{matrix}f_1 = 2(x+1)\,,\\ f_2 = 2(n-x)\,,\end{matrix}\ \text{ and }\ \underline{\theta} = \frac{x}{x+(n-x+1)v_{P_2}^2},\ \ \begin{matrix}f_1 \doteq 2(n-x+1)\,,\\ f_2 = 2x\,,\end{matrix}$$

§ 21.11, p. 697.

For $P_1 = 0.025$ and 0.005, and $P_2 = 0.975$ and 0.995 these limits have been given in Table XI.

Testing the Hypothesis $\xi_1 \le \xi_2$ Against the Alternative $\xi_1 > \xi_2$ for Two Poisson-Distributions.

The critical region at the α level of significance is

$$\frac{x_1}{x_2+1} > v_{1-\alpha}^2\big(2(x_2+1),\ 2x_1\big),\ \ \S\ 22.7,\ \text{p. 725}.$$

Tables of Fractiles of the v^2-Distribution:

R. A. FISHER and F. YATES: *Statistical Tables*, Oliver and Boyd, London, 1943. Fractiles to two decimal places for $P = 0.80,\ 0.95,\ 0.99$, and 0.999.

M. MERRINGTON and C. M. THOMPSON: *Tables of Percentage Points of the Inverted Beta (F) Distribution*, Biometrika, 33, 1943, 73–88. Fractiles to five significant figures for $P = 0.50,\ 0.75,\ 0.90,$ $0.95,\ 0.975,\ 0.99$, and 0.995.

Some Applications of the Table of the W-Distribution (Range), Table VIII, Will be Given Underneath the Table.

Applications of the Table of the One-sided Truncated Normal Distribution With Known Point of Truncation, Table IX.

The point of truncation is chosen as origin, and $x_1,\ \ldots,\ x_n$ denote the observations reckoned from this point. The standardized point of truncation is $\zeta = -\xi/\sigma$, and the degree of truncation is $\Phi(\zeta)$. We then have

$$P\{x\} = \frac{\Phi\left(\frac{x-\xi}{\sigma}\right) - \Phi(\zeta)}{1-\Phi(\zeta)}\ \ \text{ and }\ \ p\{x\} = \frac{1}{1-\Phi(\zeta)}\frac{1}{\sigma}\varphi\left(\frac{x-\xi}{\sigma}\right),\ x \ge 0\,.$$

Estimates of ξ and σ are obtained by means of Table IX as follows:

$$y = \frac{n\sum_{i=1}^{n}x_i^2}{2\left(\sum_{i=1}^{n}x_i\right)^2},\ \ z = f(y) \approx \zeta\,,\ \ s = \frac{\sum_{i=1}^{n}x_i}{n}\,g(z) \approx \sigma\,,\ \bar{x} = -zs \approx \xi\,,\ \ \S\ 6.9,\ \text{p. 144}.$$

For large values of n these estimates are approximately normally distributed with variances and covariance given by

$$\mathcal{V}\{\bar{x}\} = \frac{\sigma^2}{n}\mu_{11}(\zeta),\ \ \mathcal{V}\{s\} = \frac{\sigma^2}{n}\mu_{22}(\zeta),\ \ \mathcal{V}\{\bar{x}, s\} = \frac{\sigma^2}{n}\mu_{12}(\zeta),\ \text{and}\ \varrho\{\bar{x}, s\} = \varrho(\zeta)\,,$$

§ 9.8, p. 245, and § 11.11, p. 316.

Fractiles of s and confidence limits for σ may be found from the formulas

$$P\left\{s < \sigma\left(1 + u_P\sqrt{\frac{\mu_{22}(\zeta)}{n}}\right)\right\} \simeq P\ \ \text{and}\ \ P\left\{\frac{s}{1 + u_P\sqrt{\dfrac{\mu_{22}(\zeta)}{n}}} < \sigma\right\} \simeq P,\ \ \S\ 11.11,\ \text{p. 316}.$$

The hypothesis $\xi_1 = \xi_2$ may be tested by the u-test using

$$u \simeq \frac{\bar{x}_1 - \bar{x}_2}{s\sqrt{\left(\dfrac{1}{n_1} + \dfrac{1}{n_2}\right)\mu_{11}(z)}},\ \ \ s = \frac{n_1 s_1 + n_2 s_2}{n_1 + n_2},\ \ \S\ 15.7,\ \text{p. 410}.$$

Applications of the Table of the One-sided Censored Normal Distribution, Table X.

A sample of n observations is drawn from a normal population with parameters ξ and σ, and a number, a say, of observations are less than or equal to the known point of truncation, i. e., the values of these observations are not further specified. The sample values may be written as $(x_1, \ldots, x_{n-a}, 0, \ldots, 0)$, where the zeros symbolize the a observations less than or equal to the point of truncation which is chosen as origin. The observed degree of truncation is $h = a/n$.

Estimates of ξ and σ are obtained by means of Table X as follows:

$$y = \frac{(n-a)\sum\limits_{i=1}^{n-a} x_i^2}{2\left(\sum\limits_{i=1}^{n-a} x_i\right)^2},\ \ z = f(h, y) \approx \zeta,\ \ s = \frac{\sum\limits_{i=1}^{n-a} x_i}{n-a}\,g(h, z) \approx \sigma,\ \ \bar{x} = -zs \approx \xi,\ \ \S\ 6.9,\ \text{p. 144}.$$

The formulas for the variances and covariance of these estimates, fractiles for s, confidence limits for σ and tests of significance are the same as those given for the truncated distribution.

A derivation of the above formulas may be found in A. HALD: *Maximum Likelihood Estimation of the Parameters of a Normal Distribution Which is Truncated at a Known Point*, Skandinavisk Aktuarietidskrift, 1949, 119–134, where also the computation of Tables IX and X has been explained.

Remarks on the Table of Random Sampling Numbers, Table XIX.

The random sampling numbers have been compiled from the drawings in the Danish State Interest Lottery of 1948. Each page of 2500 numbers and the total 15,000 numbers have been tested by the frequency test, the serial test, the poker test, and the gap test as devised by M. G. KENDALL and B. BABINGTON SMITH: *Randomness and Random Sampling Numbers*, Jour. Roy. Stat. Soc., 101, 1938, 147—167. No extreme deviation from randomness was found.

TABLES

THE PROBABILITY P FOUND IN
THE TABLES OF FRACTILES IS ALWAYS THE PROBABILITY RECKONED
FROM THE L O W E R LIMIT OF THE DISTRIBUTION.

TABLE I. THE NORMAL DISTRIBUTION FUNCTION   33

$$\varphi(u) = \frac{1}{\sqrt{2\pi}} e^{-\frac{u^2}{2}} \text{ FOR } 0\cdot00 \le u \le 4\cdot99. \quad \varphi(-u) = \varphi(u).$$

u	·00	·01	·02	·03	·04	·05	·06	·07	·08	·09
·0	·3989	·3989	·3989	·3988	·3986	·3984	·3982	·3980	·3977	·3973
·1	·3970	·3965	·3961	·3956	·3951	·3945	·3939	·3932	·3925	·3918
·2	·3910	·3902	·3894	·3885	·3876	·3867	·3857	·3847	·3836	·3825
·3	·3814	·3802	·3790	·3778	·3765	·3752	·3739	·3725	·3712	·3697
·4	·3683	·3668	·3653	·3637	·3621	·3605	·3589	·3572	·3555	·3538
·5	·3521	·3503	·3485	·3467	·3448	·3429	·3410	·3391	·3372	·3352
·6	·3332	·3312	·3292	·3271	·3251	·3230	·3209	·3187	·3166	·3144
·7	·3123	·3101	·3079	·3056	·3034	·3011	·2989	·2966	·2943	·2920
·8	·2897	·2874	·2850	·2827	·2803	·2780	·2756	·2732	·2709	·2685
·9	·2661	·2637	·2613	·2589	·2565	·2541	·2516	·2492	·2468	·2444
1·0	·2420	·2396	·2371	·2347	·2323	·2299	·2275	·2251	·2227	·2203
1·1	·2179	·2155	·2131	·2107	·2083	·2059	·2036	·2012	·1989	·1965
1·2	·1942	·1919	·1895	·1872	·1849	·1826	·1804	·1781	·1758	·1736
1·3	·1714	·1691	·1669	·1647	·1626	·1604	·1582	·1561	·1539	·1518
1·4	·1497	·1476	·1456	·1435	·1415	·1394	·1374	·1354	·1334	·1315
1·5	·1295	·1276	·1257	·1238	·1219	·1200	·1182	·1163	·1145	·1127
1·6	·1109	·1092	·1074	·1057	·1040	·1023	·1006	·09893	·09728	·09566
1·7	·09405	·09246	·09089	·08933	·08780	·08628	·08478	·08329	·08183	·08038
1·8	·07895	·07754	·07614	·07477	·07341	·07206	·07074	·06943	·06814	·06687
1·9	·06562	·06438	·06316	·06195	·06077	·05959	·05844	·05730	·05618	·05508
2·0	·05399	·05292	·05186	·05082	·04980	·04879	·04780	·04682	·04586	·04491
2·1	·04398	·04307	·04217	·04128	·04041	·03955	·03871	·03788	·03706	·03626
2·2	·03547	·03470	·03394	·03319	·03246	·03174	·03103	·03034	·02965	·02898
2·3	·02833	·02768	·02705	·02643	·02582	·02522	·02463	·02406	·02349	·02294
2·4	·02239	·02186	·02134	·02083	·02033	·01984	·01936	·01888	·01842	·01797
2·5	·01753	·01709	·01667	·01625	·01585	·01545	·01506	·01468	·01431	·01394
2·6	·01358	·01323	·01289	·01256	·01223	·01191	·01160	·01130	·01100	·01071
2·7	·01042	·01014	·0^{2}9871	·0^{2}9606	·0^{2}9347	·0^{2}9094	·0^{2}8846	·0^{2}8605	·0^{2}8370	·0^{2}8140
2·8	·0^{2}7915	·0^{2}7697	·0^{2}7483	·0^{2}7274	·0^{2}7071	·0^{2}6873	·0^{2}6679	·0^{2}6491	·0^{2}6307	·0^{2}6127
2·9	·0^{2}5953	·0^{2}5782	·0^{2}5616	·0^{2}5454	·0^{2}5296	·0^{2}5143	·0^{2}4993	·0^{2}4847	·0^{2}4705	·0^{2}4567
3·0	·0^{2}4432	·0^{2}4301	·0^{2}4173	·0^{2}4049	·0^{2}3928	·0^{2}3810	·0^{2}3695	·0^{2}3584	·0^{2}3475	·0^{2}3370
3·1	·0^{2}3267	·0^{2}3167	·0^{2}3070	·0^{2}2975	·0^{2}2884	·0^{2}2794	·0^{2}2707	·0^{2}2623	·0^{2}2541	·0^{2}2461
3·2	·0^{2}2384	·0^{2}2309	·0^{2}2236	·0^{2}2165	·0^{2}2096	·0^{2}2029	·0^{2}1964	·0^{2}1901	·0^{2}1840	·0^{2}1780
3·3	·0^{2}1723	·0^{2}1667	·0^{2}1612	·0^{2}1560	·0^{2}1508	·0^{2}1459	·0^{2}1411	·0^{2}1364	·0^{2}1319	·0^{2}1275
3·4	·0^{2}1232	·0^{2}1191	·0^{2}1151	·0^{2}1112	·0^{2}1075	·0^{2}1038	·0^{2}1003	·0^{3}9689	·0^{3}9358	·0^{3}9037
3·5	·0^{3}8727	·0^{3}8426	·0^{3}8135	·0^{3}7853	·0^{3}7581	·0^{3}7317	·0^{3}7061	·0^{3}6814	·0^{3}6575	·0^{3}6343
3·6	·0^{3}6119	·0^{3}5902	·0^{3}5693	·0^{3}5490	·0^{3}5294	·0^{3}5105	·0^{3}4921	·0^{3}4744	·0^{3}4573	·0^{3}4408
3·7	·0^{3}4248	·0^{3}4093	·0^{3}3944	·0^{3}3800	·0^{3}3661	·0^{3}3526	·0^{3}3396	·0^{3}3271	·0^{3}3149	·0^{3}3032
3·8	·0^{3}2919	·0^{3}2810	·0^{3}2705	·0^{3}2604	·0^{3}2506	·0^{3}2411	·0^{3}2320	·0^{3}2232	·0^{3}2147	·0^{3}2065
3·9	·0^{3}1987	·0^{3}1910	·0^{3}1837	·0^{3}1766	·0^{3}1698	·0^{3}1633	·0^{3}1569	·0^{3}1508	·0^{3}1449	·0^{3}1393
4·0	·0^{3}1338	·0^{3}1286	·0^{3}1235	·0^{3}1186	·0^{3}1140	·0^{3}1094	·0^{3}1051	·0^{3}1009	·0^{4}9687	·0^{4}9299
4·1	·0^{4}8926	·0^{4}8567	·0^{4}8222	·0^{4}7890	·0^{4}7570	·0^{4}7263	·0^{4}6967	·0^{4}6683	·0^{4}6410	·0^{4}6147
4·2	·0^{4}5894	·0^{4}5652	·0^{4}5418	·0^{4}5194	·0^{4}4979	·0^{4}4772	·0^{4}4573	·0^{4}4382	·0^{4}4199	·0^{4}4023
4·3	·0^{4}3854	·0^{4}3691	·0^{4}3535	·0^{4}3386	·0^{4}3242	·0^{4}3104	·0^{4}2972	·0^{4}2845	·0^{4}2723	·0^{4}2606
4·4	·0^{4}2494	·0^{4}2387	·0^{4}2284	·0^{4}2185	·0^{4}2090	·0^{4}1999	·0^{4}1912	·0^{4}1829	·0^{4}1749	·0^{4}1672
4·5	·0^{4}1598	·0^{4}1528	·0^{4}1461	·0^{4}1396	·0^{4}1334	·0^{4}1275	·0^{4}1218	·0^{4}1164	·0^{4}1112	·0^{4}1062
4·6	·0^{4}1014	·0^{5}9684	·0^{5}9248	·0^{5}8830	·0^{5}8430	·0^{5}8047	·0^{5}7681	·0^{5}7331	·0^{5}6996	·0^{5}6676
4·7	·0^{5}6370	·0^{5}6077	·0^{5}5797	·0^{5}5530	·0^{5}5274	·0^{5}5030	·0^{5}4796	·0^{5}4573	·0^{5}4360	·0^{5}4156
4·8	·0^{5}3961	·0^{5}3775	·0^{5}3598	·0^{5}3428	·0^{5}3267	·0^{5}3112	·0^{5}2965	·0^{5}2824	·0^{5}2690	·0^{5}2561
4·9	·0^{5}2439	·0^{5}2322	·0^{5}2211	·0^{5}2105	·0^{5}2003	·0^{5}1907	·0^{5}1814	·0^{5}1727	·0^{5}1643	·0^{5}1563

Example: $\varphi(3\cdot57) = \varphi(-3\cdot57) = \cdot0^3 6814 = 0\cdot0006814.$

TABLE II. THE CUMULATIVE NORMAL DISTRIBUTION FUNCTION

$$\Phi(u) = \frac{1}{\sqrt{2\pi}} \int_{-\infty}^{u} e^{-\frac{x^2}{2}}\, dx \quad \text{FOR} \quad -4{\cdot}99 \leq u \leq 0{\cdot}00.$$

u	·00	·01	·02	·03	·04	·05	·06	·07	·08	·09
− ·0	·5000	·4960	·4920	·4880	·4840	·4801	·4761	·4721	·4681	·4641
− ·1	·4602	·4562	·4522	·4483	·4443	·4404	·4364	·4325	·4286	·4247
− ·2	·4207	·4168	·4129	·4090	·4052	·4013	·3974	·3936	·3897	·3859
− ·3	·3821	·3783	·3745	·3707	·3669	·3632	·3594	·3557	·3520	·3483
− ·4	·3446	·3409	·3372	·3336	·3300	·3264	·3228	·3192	·3156	·3121
− ·5	·3085	·3050	·3015	·2981	·2946	·2912	·2877	·2843	·2810	·2776
− ·6	·2743	·2709	·2676	·2643	·2611	·2578	·2546	·2514	·2483	·2451
− ·7	·2420	·2389	·2358	·2327	·2297	·2266	·2236	·2206	·2177	·2148
− ·8	·2119	·2090	·2061	·2033	·2005	·1977	·1949	·1922	·1894	·1867
− ·9	·1841	·1814	·1788	·1762	·1736	·1711	·1685	·1660	·1635	·1611
−1·0	·1587	·1562	·1539	·1515	·1492	·1469	·1446	·1423	·1401	·1379
−1·1	·1357	·1335	·1314	·1292	·1271	·1251	·1230	·1210	·1190	·1170
−1·2	·1151	·1131	·1112	·1093	·1075	·1056	·1038	·1020	·1003	·09853
−1·3	·09680	·09510	·09342	·09176	·09012	·08851	·08691	·08534	·08379	·08226
−1·4	·08076	·07927	·07780	·07636	·07493	·07353	·07215	·07078	·06944	·06811
−1·5	·06681	·06552	·06426	·06301	·06178	·06057	·05938	·05821	·05705	·05592
−1·6	·05480	·05370	·05262	·05155	·05050	·04947	·04846	·04746	·04648	·04551
−1·7	·04457	·04363	·04272	·04182	·04093	·04006	·03920	·03836	·03754	·03673
−1·8	·03593	·03515	·03438	·03362	·03288	·03216	·03144	·03074	·03005	·02938
−1·9	·02872	·02807	·02743	·02680	·02619	·02559	·02500	·02442	·02385	·02330
−2·0	·02275	·02222	·02169	·02118	·02068	·02018	·01970	·01923	·01876	·01831
−2·1	·01786	·01743	·01700	·01659	·01618	·01578	·01539	·01500	·01463	·01426
−2·2	·01390	·01355	·01321	·01287	·01255	·01222	·01191	·01160	·01130	·01101
−2·3	·01072	·01044	·01017	$\cdot0^2 9903$	$\cdot0^2 9642$	$\cdot0^2 9387$	$\cdot0^2 9137$	$\cdot0^2 8894$	$\cdot0^2 8656$	$\cdot0^2 8424$
−2·4	$\cdot0^2 8198$	$\cdot0^2 7976$	$\cdot0^2 7760$	$\cdot0^2 7549$	$\cdot0^2 7344$	$\cdot0^2 7143$	$\cdot0^2 6947$	$\cdot0^2 6756$	$\cdot0^2 6569$	$\cdot0^2 6387$
−2·5	$\cdot0^2 6210$	$\cdot0^2 6037$	$\cdot0^2 5868$	$\cdot0^2 5703$	$\cdot0^2 5543$	$\cdot0^2 5386$	$\cdot0^2 5234$	$\cdot0^2 5085$	$\cdot0^2 4940$	$\cdot0^2 4799$
−2·6	$\cdot0^2 4661$	$\cdot0^2 4527$	$\cdot0^2 4396$	$\cdot0^2 4269$	$\cdot0^2 4145$	$\cdot0^2 4025$	$\cdot0^2 3907$	$\cdot0^2 3793$	$\cdot0^2 3681$	$\cdot0^2 3573$
−2·7	$\cdot0^2 3467$	$\cdot0^2 3364$	$\cdot0^2 3264$	$\cdot0^2 3167$	$\cdot0^2 3072$	$\cdot0^2 2980$	$\cdot0^2 2890$	$\cdot0^2 2803$	$\cdot0^2 2718$	$\cdot0^2 2635$
−2·8	$\cdot0^2 2555$	$\cdot0^2 2477$	$\cdot0^2 2401$	$\cdot0^2 2327$	$\cdot0^2 2256$	$\cdot0^2 2186$	$\cdot0^2 2118$	$\cdot0^2 2052$	$\cdot0^2 1988$	$\cdot0^2 1926$
−2·9	$\cdot0^2 1866$	$\cdot0^2 1807$	$\cdot0^2 1750$	$\cdot0^2 1695$	$\cdot0^2 1641$	$\cdot0^2 1589$	$\cdot0^2 1538$	$\cdot0^2 1489$	$\cdot0^2 1441$	$\cdot0^2 1395$
−3·0	$\cdot0^2 1350$	$\cdot0^2 1306$	$\cdot0^2 1264$	$\cdot0^2 1223$	$\cdot0^2 1183$	$\cdot0^2 1144$	$\cdot0^2 1107$	$\cdot0^2 1070$	$\cdot0^2 1035$	$\cdot0^2 1001$
−3·1	$\cdot0^3 9676$	$\cdot0^3 9354$	$\cdot0^3 9043$	$\cdot0^3 8740$	$\cdot0^3 8447$	$\cdot0^3 8164$	$\cdot0^3 7888$	$\cdot0^3 7622$	$\cdot0^3 7364$	$\cdot0^3 7114$
−3·2	$\cdot0^3 6871$	$\cdot0^3 6637$	$\cdot0^3 6410$	$\cdot0^3 6190$	$\cdot0^3 5976$	$\cdot0^3 5770$	$\cdot0^3 5571$	$\cdot0^3 5377$	$\cdot0^3 5190$	$\cdot0^3 5009$
−3·3	$\cdot0^3 4834$	$\cdot0^3 4665$	$\cdot0^3 4501$	$\cdot0^3 4342$	$\cdot0^3 4189$	$\cdot0^3 4041$	$\cdot0^3 3897$	$\cdot0^3 3758$	$\cdot0^3 3624$	$\cdot0^3 3495$
−3·4	$\cdot0^3 3369$	$\cdot0^3 3248$	$\cdot0^3 3131$	$\cdot0^3 3018$	$\cdot0^3 2909$	$\cdot0^3 2803$	$\cdot0^3 2701$	$\cdot0^3 2602$	$\cdot0^3 2507$	$\cdot0^3 2415$
−3·5	$\cdot0^3 2326$	$\cdot0^3 2241$	$\cdot0^3 2158$	$\cdot0^3 2078$	$\cdot0^3 2001$	$\cdot0^3 1926$	$\cdot0^3 1854$	$\cdot0^3 1785$	$\cdot0^3 1718$	$\cdot0^3 1653$
−3·6	$\cdot0^3 1591$	$\cdot0^3 1531$	$\cdot0^3 1473$	$\cdot0^3 1417$	$\cdot0^3 1363$	$\cdot0^3 1311$	$\cdot0^3 1261$	$\cdot0^3 1213$	$\cdot0^3 1166$	$\cdot0^3 1121$
−3·7	$\cdot0^3 1078$	$\cdot0^3 1036$	$\cdot0^4 9961$	$\cdot0^4 9574$	$\cdot0^4 9201$	$\cdot0^4 8842$	$\cdot0^4 8496$	$\cdot0^4 8162$	$\cdot0^4 7841$	$\cdot0^4 7532$
−3·8	$\cdot0^4 7235$	$\cdot0^4 6948$	$\cdot0^4 6673$	$\cdot0^4 6407$	$\cdot0^4 6152$	$\cdot0^4 5906$	$\cdot0^4 5669$	$\cdot0^4 5442$	$\cdot0^4 5223$	$\cdot0^4 5012$
−3·9	$\cdot0^4 4810$	$\cdot0^4 4615$	$\cdot0^4 4427$	$\cdot0^4 4247$	$\cdot0^4 4074$	$\cdot0^4 3908$	$\cdot0^4 3747$	$\cdot0^4 3594$	$\cdot0^4 3446$	$\cdot0^4 3304$
−4·0	$\cdot0^4 3167$	$\cdot0^4 3036$	$\cdot0^4 2910$	$\cdot0^4 2789$	$\cdot0^4 2673$	$\cdot0^4 2561$	$\cdot0^4 2454$	$\cdot0^4 2351$	$\cdot0^4 2252$	$\cdot0^4 2157$
−4·1	$\cdot0^4 2066$	$\cdot0^4 1978$	$\cdot0^4 1894$	$\cdot0^4 1814$	$\cdot0^4 1737$	$\cdot0^4 1662$	$\cdot0^4 1591$	$\cdot0^4 1523$	$\cdot0^4 1458$	$\cdot0^4 1395$
−4·2	$\cdot0^4 1335$	$\cdot0^4 1277$	$\cdot0^4 1222$	$\cdot0^4 1168$	$\cdot0^4 1118$	$\cdot0^4 1069$	$\cdot0^4 1022$	$\cdot0^5 9774$	$\cdot0^5 9345$	$\cdot0^5 8934$
−4·3	$\cdot0^5 8540$	$\cdot0^5 8163$	$\cdot0^5 7801$	$\cdot0^5 7455$	$\cdot0^5 7124$	$\cdot0^5 6807$	$\cdot0^5 6503$	$\cdot0^5 6212$	$\cdot0^5 5934$	$\cdot0^5 5668$
−4·4	$\cdot0^5 5413$	$\cdot0^5 5169$	$\cdot0^5 4935$	$\cdot0^5 4712$	$\cdot0^5 4498$	$\cdot0^5 4294$	$\cdot0^5 4098$	$\cdot0^5 3911$	$\cdot0^5 3732$	$\cdot0^5 3561$
−4·5	$\cdot0^5 3398$	$\cdot0^5 3241$	$\cdot0^5 3092$	$\cdot0^5 2949$	$\cdot0^5 2813$	$\cdot0^5 2682$	$\cdot0^5 2558$	$\cdot0^5 2439$	$\cdot0^5 2325$	$\cdot0^5 2216$
−4·6	$\cdot0^5 2112$	$\cdot0^5 2013$	$\cdot0^5 1919$	$\cdot0^5 1828$	$\cdot0^5 1742$	$\cdot0^5 1660$	$\cdot0^5 1581$	$\cdot0^5 1506$	$\cdot0^5 1434$	$\cdot0^5 1366$
−4·7	$\cdot0^5 1301$	$\cdot0^5 1239$	$\cdot0^5 1179$	$\cdot0^5 1123$	$\cdot0^5 1069$	$\cdot0^5 1017$	$\cdot0^6 9680$	$\cdot0^6 9211$	$\cdot0^6 8765$	$\cdot0^6 8339$
−4·8	$\cdot0^6 7933$	$\cdot0^6 7547$	$\cdot0^6 7178$	$\cdot0^6 6827$	$\cdot0^6 6492$	$\cdot0^6 6173$	$\cdot0^6 5869$	$\cdot0^6 5580$	$\cdot0^6 5304$	$\cdot0^6 5042$
−4·9	$\cdot0^6 4792$	$\cdot0^6 4554$	$\cdot0^6 4327$	$\cdot0^6 4111$	$\cdot0^6 3906$	$\cdot0^6 3711$	$\cdot0^6 3525$	$\cdot0^6 3348$	$\cdot0^6 3179$	$\cdot0^6 3019$

Example: $\Phi(-3{\cdot}57) = \cdot0^3 1785 = 0{\cdot}0001785.$

TABLE II. THE CUMULATIVE NORMAL DISTRIBUTION FUNCTION   35

$$\Phi(u) = \frac{1}{\sqrt{2\pi}} \int_{-\infty}^{u} e^{-\frac{x^2}{2}} dx \quad \text{FOR} \quad 0\cdot00 \leq u \leq 4\cdot99.$$

u	·00	·01	·02	·03	·04	·05	·06	·07	·08	·09
·0	·5000	·5040	·5080	·5120	·5160	·5199	·5239	·5279	·5319	·5359
·1	·5398	·5438	·5478	·5517	·5557	·5596	·5636	·5675	·5714	·5753
·2	·5793	·5832	·5871	·5910	·5948	·5987	·6026	·6064	·6103	·6141
·3	·6179	·6217	·6255	·6293	·6331	·6368	·6406	·6443	·6480	·6517
·4	·6554	·6591	·6628	·6664	·6700	·6736	·6772	·6808	·6844	·6879
·5	·6915	·6950	·6985	·7019	·7054	·7088	·7123	·7157	·7190	·7224
·6	·7257	·7291	·7324	·7357	·7389	·7422	·7454	·7486	·7517	·7549
·7	·7580	·7611	·7642	·7673	·7703	·7734	·7764	·7794	·7823	·7852
·8	·7881	·7910	·7939	·7967	·7995	·8023	·8051	·8078	·8106	·8133
·9	·8159	·8186	·8212	·8238	·8264	·8289	·8315	·8340	·8365	·8389
1·0	·8413	·8438	·8461	·8485	·8508	·8531	·8554	·8577	·8599	·8621
1·1	·8643	·8665	·8686	·8708	·8729	·8749	·8770	·8790	·8810	·8830
1·2	·8849	·8869	·8888	·8907	·8925	·8944	·8962	·8980	·8997	·90147
1·3	·90320	·90490	·90658	·90824	·90988	·91149	·91309	·91466	·91621	·91774
1·4	·91924	·92073	·92220	·92364	·92507	·92647	·92785	·92922	·93056	·93189
1·5	·93319	·93448	·93574	·93699	·93822	·93943	·94062	·94179	·94295	·94408
1·6	·94520	·94630	·94738	·94845	·94950	·95053	·95154	·95254	·95352	·95449
1·7	·95543	·95637	·95728	·95818	·95907	·95994	·96080	·96164	·96246	·96327
1·8	·96407	·96485	·96562	·96638	·96712	·96784	·96856	·96926	·96995	·97062
1·9	·97128	·97193	·97257	·97320	·97381	·97441	·97500	·97558	·97615	·97670
2·0	·97725	·97778	·97831	·97882	·97932	·97982	·98030	·98077	·98124	·98169
2·1	·98214	·98257	·98300	·98341	·98382	·98422	·98461	·98500	·98537	·98574
2·2	·98610	·98645	·98679	·98713	·98745	·98778	·98809	·98840	·98870	·98899
2·3	·98928	·98956	·98983	$\cdot9^2 0097$	$\cdot9^2 0358$	$\cdot9^2 0613$	$\cdot9^2 0863$	$\cdot9^2 1106$	$\cdot9^2 1344$	$\cdot9^2 1576$
2·4	$\cdot9^2 1802$	$\cdot9^2 2024$	$\cdot9^2 2240$	$\cdot9^2 2451$	$\cdot9^2 2656$	$\cdot9^2 2857$	$\cdot9^2 3053$	$\cdot9^2 3244$	$\cdot9^2 3431$	$\cdot9^2 3613$
2·5	$\cdot9^2 3790$	$\cdot9^2 3963$	$\cdot9^2 4132$	$\cdot9^2 4297$	$\cdot9^2 4457$	$\cdot9^2 4614$	$\cdot9^2 4766$	$\cdot9^2 4915$	$\cdot9^2 5060$	$\cdot9^2 5201$
2·6	$\cdot9^2 5339$	$\cdot9^2 5473$	$\cdot9^2 5604$	$\cdot9^2 5731$	$\cdot9^2 5855$	$\cdot9^2 5975$	$\cdot9^2 6093$	$\cdot9^2 6207$	$\cdot9^2 6319$	$\cdot9^2 6427$
2·7	$\cdot9^2 6533$	$\cdot9^2 6636$	$\cdot9^2 6736$	$\cdot9^2 6833$	$\cdot9^2 6928$	$\cdot9^2 7020$	$\cdot9^2 7110$	$\cdot9^2 7197$	$\cdot9^2 7282$	$\cdot9^2 7365$
2·8	$\cdot9^2 7445$	$\cdot9^2 7523$	$\cdot9^2 7599$	$\cdot9^2 7673$	$\cdot9^2 7744$	$\cdot9^2 7814$	$\cdot9^2 7882$	$\cdot9^2 7948$	$\cdot9^2 8012$	$\cdot9^2 8074$
2·9	$\cdot9^2 8134$	$\cdot9^2 8193$	$\cdot9^2 8250$	$\cdot9^2 8305$	$\cdot9^2 8359$	$\cdot9^2 8411$	$\cdot9^2 8462$	$\cdot9^2 8511$	$\cdot9^2 8559$	$\cdot9^2 8605$
3·0	$\cdot9^2 8650$	$\cdot9^2 8694$	$\cdot9^2 8736$	$\cdot9^2 8777$	$\cdot9^2 8817$	$\cdot9^2 8856$	$\cdot9^2 8893$	$\cdot9^2 8930$	$\cdot9^2 8965$	$\cdot9^2 8999$
3·1	$\cdot9^3 0324$	$\cdot9^3 0646$	$\cdot9^3 0957$	$\cdot9^3 1260$	$\cdot9^3 1553$	$\cdot9^3 1836$	$\cdot9^3 2112$	$\cdot9^3 2378$	$\cdot9^3 2636$	$\cdot9^3 2886$
3·2	$\cdot9^3 3129$	$\cdot9^3 3363$	$\cdot9^3 3590$	$\cdot9^3 3810$	$\cdot9^3 4024$	$\cdot9^3 4230$	$\cdot9^3 4429$	$\cdot9^3 4623$	$\cdot9^3 4810$	$\cdot9^3 4991$
3·3	$\cdot9^3 5166$	$\cdot9^3 5335$	$\cdot9^3 5499$	$\cdot9^3 5658$	$\cdot9^3 5811$	$\cdot9^3 5959$	$\cdot9^3 6103$	$\cdot9^3 6242$	$\cdot9^3 6376$	$\cdot9^3 6505$
3·4	$\cdot9^3 6631$	$\cdot9^3 6752$	$\cdot9^3 6869$	$\cdot9^3 6982$	$\cdot9^3 7091$	$\cdot9^3 7197$	$\cdot9^3 7299$	$\cdot9^3 7398$	$\cdot9^3 7493$	$\cdot9^3 7585$
3·5	$\cdot9^3 7674$	$\cdot9^3 7759$	$\cdot9^3 7842$	$\cdot9^3 7922$	$\cdot9^3 7999$	$\cdot9^3 8074$	$\cdot9^3 8146$	$\cdot9^3 8215$	$\cdot9^3 8282$	$\cdot9^3 8347$
3·6	$\cdot9^3 8409$	$\cdot9^3 8469$	$\cdot9^3 8527$	$\cdot9^3 8583$	$\cdot9^3 8637$	$\cdot9^3 8689$	$\cdot9^3 8739$	$\cdot9^3 8787$	$\cdot9^3 8834$	$\cdot9^3 8879$
3·7	$\cdot9^3 8922$	$\cdot9^3 8964$	$\cdot9^4 0039$	$\cdot9^4 0426$	$\cdot9^4 0799$	$\cdot9^4 1158$	$\cdot9^4 1504$	$\cdot9^4 1838$	$\cdot9^4 2159$	$\cdot9^4 2468$
3·8	$\cdot9^4 2765$	$\cdot9^4 3052$	$\cdot9^4 3327$	$\cdot9^4 3593$	$\cdot9^4 3848$	$\cdot9^4 4094$	$\cdot9^4 4331$	$\cdot9^4 4558$	$\cdot9^4 4777$	$\cdot9^4 4988$
3·9	$\cdot9^4 5190$	$\cdot9^4 5385$	$\cdot9^4 5573$	$\cdot9^4 5753$	$\cdot9^4 5926$	$\cdot9^4 6092$	$\cdot9^4 6253$	$\cdot9^4 6406$	$\cdot9^4 6554$	$\cdot9^4 6696$
4·0	$\cdot9^4 6833$	$\cdot9^4 6964$	$\cdot9^4 7090$	$\cdot9^4 7211$	$\cdot9^4 7327$	$\cdot9^4 7439$	$\cdot9^4 7546$	$\cdot9^4 7649$	$\cdot9^4 7748$	$\cdot9^4 7843$
4·1	$\cdot9^4 7934$	$\cdot9^4 8022$	$\cdot9^4 8106$	$\cdot9^4 8186$	$\cdot9^4 8263$	$\cdot9^4 8338$	$\cdot9^4 8409$	$\cdot9^4 8477$	$\cdot9^4 8542$	$\cdot9^4 8605$
4·2	$\cdot9^4 8665$	$\cdot9^4 8723$	$\cdot9^4 8778$	$\cdot9^4 8832$	$\cdot9^4 8882$	$\cdot9^4 8931$	$\cdot9^4 8978$	$\cdot9^5 0226$	$\cdot9^5 0655$	$\cdot9^5 1066$
4·3	$\cdot9^5 1460$	$\cdot9^5 1837$	$\cdot9^5 2199$	$\cdot9^5 2545$	$\cdot9^5 2876$	$\cdot9^5 3193$	$\cdot9^5 3497$	$\cdot9^5 3788$	$\cdot9^5 4066$	$\cdot9^5 4332$
4·4	$\cdot9^5 4587$	$\cdot9^5 4831$	$\cdot9^5 5065$	$\cdot9^5 5288$	$\cdot9^5 5502$	$\cdot9^5 5706$	$\cdot9^5 5902$	$\cdot9^5 6089$	$\cdot9^5 6268$	$\cdot9^5 6439$
4·5	$\cdot9^5 6602$	$\cdot9^5 6759$	$\cdot9^5 6908$	$\cdot9^5 7051$	$\cdot9^5 7187$	$\cdot9^5 7318$	$\cdot9^5 7442$	$\cdot9^5 7561$	$\cdot9^5 7675$	$\cdot9^5 7784$
4·6	$\cdot9^5 7888$	$\cdot9^5 7987$	$\cdot9^5 8081$	$\cdot9^5 8172$	$\cdot9^5 8258$	$\cdot9^5 8340$	$\cdot9^5 8419$	$\cdot9^5 8494$	$\cdot9^5 8566$	$\cdot9^5 8634$
4·7	$\cdot9^5 8699$	$\cdot9^5 8761$	$\cdot9^5 8821$	$\cdot9^5 8877$	$\cdot9^5 8931$	$\cdot9^5 8983$	$\cdot9^6 0320$	$\cdot9^6 0789$	$\cdot9^6 1235$	$\cdot9^6 1661$
4·8	$\cdot9^6 2067$	$\cdot9^6 2453$	$\cdot9^6 2822$	$\cdot9^6 3173$	$\cdot9^6 3508$	$\cdot9^6 3827$	$\cdot9^6 4131$	$\cdot9^6 4420$	$\cdot9^6 4696$	$\cdot9^6 4958$
4·9	$\cdot9^6 5208$	$\cdot9^6 5446$	$\cdot9^6 5673$	$\cdot9^6 5889$	$\cdot9^6 6094$	$\cdot9^6 6289$	$\cdot9^6 6475$	$\cdot9^6 6652$	$\cdot9^6 6821$	$\cdot9^6 6981$

Example: $\Phi(3\cdot57) = \cdot9^3 8215 = 0\cdot9998215.$

TABLE III. FRACTILES+5 (PROBITS) OF THE NORMAL DISTRIBUTION

TABLE OF u_P+5 FOR $\cdot000 \leq P \leq \cdot499$.

100 P	·0	·1	·2	·3	·4	·5	·6	·7	·8	·9
0	—	1·910	2·122	2·252	2·348	2·424	2·488	2·543	2·591	2·634
1	2·674	2·710	2·743	2·774	2·803	2·830	2·856	2·880	2·903	2·925
2	2·946	2·966	2·986	3·005	3·023	3·040	3·057	3·073	3·089	3·104
3	3·119	3·134	3·148	3·162	3·175	3·188	3·201	3·213	3·226	3·238
4	3·249	3·261	3·272	3·283	3·294	3·305	3·315	3·325	3·335	3·345
5	3·355	3·365	3·374	3·384	3·393	3·402	3·411	3·420	3·428	3·437
6	3·445	3·454	3·462	3·470	3·478	3·486	3·494	3·501	3·509	3·517
7	3·524	3·532	3·539	3·546	3·553	3·560	3·567	3·574	3·581	3·588
8	3·595	3·602	3·608	3·615	3·621	3·628	3·634	3·641	3·647	3·653
9	3·659	3·665	3·671	3·677	3·683	3·689	3·695	3·701	3·707	3·713
10	3·718	3·724	3·730	3·735	3·741	3·746	3·752	3·757	3·763	3·768
11	3·773	3·779	3·784	3·789	3·794	3·800	3·805	3·810	3·815	3·820
12	3·825	3·830	3·835	3·840	3·845	3·850	3·854	3·859	3·864	3·869
13	3·874	3·878	3·883	3·888	3·892	3·897	3·902	3·906	3·911	3·915
14	3·920	3·924	3·929	3·933	3·937	3·942	3·946	3·951	3·955	3·959
15	3·964	3·968	3·972	3·976	3·981	3·985	3·989	3·993	3·997	4·001
16	4·006	4·010	4·014	4·018	4·022	4·026	4·030	4·034	4·038	4·042
17	4·046	4·050	4·054	4·058	4·062	4·065	4·069	4·073	4·077	4·081
18	4·085	4·088	4·092	4·096	4·100	4·104	4·107	4·111	4·115	4·118
19	4·122	4·126	4·129	4·133	4·137	4·140	4·144	4·148	4·151	4·155
20	4·158	4·162	4·166	4·169	4·173	4·176	4·180	4·183	4·187	4·190
21	4·194	4·197	4·200	4·204	4·207	4·211	4·214	4·218	4·221	4·224
22	4·228	4·231	4·235	4·238	4·241	4·245	4·248	4·251	4·255	4·258
23	4·261	4·264	4·268	4·271	4·274	4·278	4·281	4·284	4·287	4·290
24	4·294	4·297	4·300	4·303	4·307	4·310	4·313	4·316	4·319	4·322
25	4·326	4·329	4·332	4·335	4·338	4·341	4·344	4·347	4·350	4·354
26	4·357	4·360	4·363	4·366	4·369	4·372	4·375	4·378	4·381	4·384
27	4·387	4·390	4·393	4·396	4·399	4·402	4·405	4·408	4·411	4·414
28	4·417	4·420	4·423	4·426	4·429	4·432	4·435	4·438	4·441	4·444
29	4·447	4·450	4·452	4·455	4·458	4·461	4·464	4·467	4·470	4·473
30	4·476	4·478	4·481	4·484	4·487	4·490	4·493	4·496	4·498	4·501
31	4·504	4·507	4·510	4·513	4·515	4·518	4·521	4·524	4·527	4·530
32	4·532	4·535	4·538	4·541	4·543	4·546	4·549	4·552	4·555	4·557
33	4·560	4·563	4·566	4·568	4·571	4·574	4·577	4·579	4·582	4·585
34	4·588	4·590	4·593	4·596	4·598	4·601	4·604	4·607	4·609	4·612
35	4·615	4·617	4·620	4·623	4·625	4·628	4·631	4·634	4·636	4·639
36	4·642	4·644	4·647	4·650	4·652	4·655	4·658	4·660	4·663	4·665
37	4·668	4·671	4·673	4·676	4·679	4·681	4·684	4·687	4·689	4·692
38	4·695	4·697	4·700	4·702	4·705	4·708	4·710	4·713	4·715	4·718
39	4·721	4·723	4·726	4·728	4·731	4·734	4·736	4·739	4·741	4·744
40	4·747	4·749	4·752	4·754	4·757	4·760	4·762	4·765	4·767	4·770
41	4·772	4·775	4·778	4·780	4·783	4·785	4·788	4·790	4·793	4·796
42	4·798	4·801	4·803	4·806	4·808	4·811	4·813	4·816	4·819	4·821
43	4·824	4·826	4·829	4·831	4·834	4·836	4·839	4·841	4·844	4·846
44	4·849	4·852	4·854	4·857	4·859	4·862	4·864	4·867	4·869	4·872
45	4·874	4·877	4·879	4·882	4·884	4·887	4·889	4·892	4·895	4·897
46	4·900	4·902	4·905	4·907	4·910	4·912	4·915	4·917	4·920	4·922
47	4·925	4·927	4·930	4·932	4·935	4·937	4·940	4·942	4·945	4·947
48	4·950	4·952	4·955	4·957	4·960	4·962	4·965	4·967	4·970	4·972
49	4·975	4·977	4·980	4·982	4·985	4·987	4·990	4·992	4·995	4·997

Example: $u_{\cdot025}+5\cdot000 = 3\cdot040$.

This table has been abridged from Table IX of R. A. FISHER and F. YATES: *Statistical*

TABLE III. FRACTILES+5 (PROBITS) OF THE NORMAL DISTRIBUTION

TABLE OF u_P+5 FOR $\cdot500 \leq P \leq \cdot999$.

100 P	·0	·1	·2	·3	·4	·5	·6	·7	·8	·9
50	5·000	5·003	5·005	5·008	5·010	5·013	5·015	5·018	5·020	5·023
51	5·025	5·028	5·030	5·033	5·035	5·038	5·040	5·043	5·045	5·048
52	5·050	5·053	5·055	5·058	5·060	5·063	5·065	5·068	5·070	5·073
53	5·075	5·078	5·080	5·083	5·085	5·088	5·090	5·093	5·095	5·098
54	5·100	5·103	5·105	5·108	5·111	5·113	5·116	5·118	5·121	5·123
55	5·126	5·128	5·131	5·133	5·136	5·138	5·141	5·143	5·146	5·148
56	5·151	5·154	5·156	5·159	5·161	5·164	5·166	5·169	5·171	5·174
57	5·176	5·179	5·181	5·184	5·187	5·189	5·192	5·194	5·197	5·199
58	5·202	5·204	5·207	5·210	5·212	5·215	5·217	5·220	5·222	5·225
59	5·228	5·230	5·233	5·235	5·238	5·240	5·243	5·246	5·248	5·251
60	5·253	5·256	5·259	5·261	5·264	5·266	5·269	5·272	5·274	5·277
61	5·279	5·282	5·285	5·287	5·290	5·292	5·295	5·298	5·300	5·303
62	5·305	5·308	5·311	5·313	5·316	5·319	5·321	5·324	5·327	5·329
63	5·332	5·335	5·337	5·340	5·342	5·345	5·348	5·350	5·353	5·356
64	5·358	5·361	5·364	5·366	5·369	5·372	5·375	5·377	5·380	5·383
65	5·385	5·388	5·391	5·393	5·396	5·399	5·402	5·404	5·407	5·410
66	5·412	5·415	5·418	5·421	5·423	5·426	5·429	5·432	5·434	5·437
67	5·440	5·443	5·445	5·448	5·451	5·454	5·457	5·459	5·462	5·465
68	5·468	5·470	5·473	5·476	5·479	5·482	5·485	5·487	5·490	5·493
69	5·496	5·499	5·502	5·504	5·507	5·510	5·513	5·516	5·519	5·522
70	5·524	5·527	5·530	5·533	5·536	5·539	5·542	5·545	5·548	5·550
71	5·553	5·556	5·559	5·562	5·565	5·568	5·571	5·574	5·577	5·580
72	5·583	5·586	5·589	5·592	5·595	5·598	5·601	5·604	5·607	5·610
73	5·613	5·616	5·619	5·622	5·625	5·628	5·631	5·634	5·637	5·640
74	5·643	5·646	5·650	5·653	5·656	5·659	5·662	5·665	5·668	5·671
75	5·674	5·678	5·681	5·684	5·687	5·690	5·693	5·697	5·700	5·703
76	5·706	5·710	5·713	5·716	5·719	5·722	5·726	5·729	5·732	5·736
77	5·739	5·742	5·745	5·749	5·752	5·755	5·759	5·762	5·765	5·769
78	5·772	5·776	5·779	5·782	5·786	5·789	5·793	5·796	5·800	5·803
79	5·806	5·810	5·813	5·817	5·820	5·824	5·827	5·831	5·834	5·838
80	5·842	5·845	5·849	5·852	5·856	5·860	5·863	5·867	5·871	5·874
81	5·878	5·882	5·885	5·889	5·893	5·896	5·900	5·904	5·908	5·912
82	5·915	5·919	5·923	5·927	5·931	5·935	5·938	5·942	5·946	5·950
83	5·954	5·958	5·962	5·966	5·970	5·974	5·978	5·982	5·986	5·990
84	5·994	5·999	6·003	6·007	6·011	6·015	6·019	6·024	6·028	6·032
85	6·036	6·041	6·045	6·049	6·054	6·058	6·063	6·067	6·071	6·076
86	6·080	6·085	6·089	6·094	6·098	6·103	6·108	6·112	6·117	6·122
87	6·126	6·131	6·136	6·141	6·146	6·150	6·155	6·160	6·165	6·170
88	6·175	6·180	6·185	6·190	6·195	6·200	6·206	6·211	6·216	6·221
89	6·227	6·232	6·237	6·243	6·248	6·254	6·259	6·265	6·270	6·276
90	6·282	6·287	6·293	6·299	6·305	6·311	6·317	6·323	6·329	6·335
91	6·341	6·347	6·353	6·359	6·366	6·372	6·379	6·385	6·392	6·398
92	6·405	6·412	6·419	6·426	6·433	6·440	6·447	6·454	6·461	6·468
93	6·476	6·483	6·491	6·499	6·506	6·514	6·522	6·530	6·538	6·546
94	6·555	6·563	6·572	6·580	6·589	6·598	6·607	6·616	6·626	6·635
95	6·645	6·655	6·665	6·675	6·685	6·695	6·706	6·717	6·728	6·739
96	6·751	6·762	6·774	6·787	6·799	6·812	6·825	6·838	6·852	6·866
97	6·881	6·896	6·911	6·927	6·943	6·960	6·977	6·995	7·014	7·034
98	7·054	7·075	7·097	7·120	7·144	7·170	7·197	7·226	7·257	7·290
99	7·326	7·366	7·409	7·457	7·512	7·576	7·652	7·748	7·878	8·090

Example: $u_{\cdot975}+5\cdot000 = 6\cdot960$.

and Boyd, Edinburgh, by permission of the authors and the publishers.

TABLE III. FRACTILES+5 (PROBITS) OF THE NORMAL DISTRIBUTION

Table of u_P+5 for $\cdot0000 \leq P \leq \cdot0249$ and $\cdot9750 \leq P \leq \cdot9999$.

100 P	·00	·01	·02	·03	·04	·05	·06	·07	·08	·09
0·0	—.	1·281	1·460	1·568	1·647	1·709	1·761	1·805	1·844	1·879
0·1	1·910	1·938	1·964	1·989	2·011	2·032	2·052	2·071	2·089	2·106
0·2	2·122	2·137	2·152	2·166	2·180	2·193	2·206	2·218	2·230	2·241
0·3	2·252	2·263	2·273	2·284	2·294	2·303	2·313	2·322	2·331	2·339
0·4	2·348	2·356	2·364	2·372	2·380	2·388	2·395	2·403	2·410	2·417
0·5	2·424	2·431	2·438	2·444	2·451	2·457	2·464	2·470	2·476	2·482
0·6	2·488	2·494	2·499	2·505	2·511	2·516	2·522	2·527	2·532	2·538
0·7	2·543	2·548	2·553	2·558	2·563	2·568	2·572	2·577	2·582	2·586
0·8	2·591	2·596	2·600	2·605	2·609	2·613	2·618	2·622	2·626	2·630
0·9	2·634	2·638	2·643	2·647	2·651	2·654	2·658	2·662	2·666	2·670
1·0	2·674	2·677	2·681	2·685	2·688	2·692	2·696	2·699	2·703	2·706
1·1	2·710	2·713	2·716	2·720	2·723	2·727	2·730	2·733	2·736	2·740
1·2	2·743	2·746	2·749	2·752	2·755	2·759	2·762	2·765	2·768	2·771
1·3	2·774	2·777	2·780	2·783	2·786	2·788	2·791	2·794	2·797	2·800
1·4	2·803	2·806	2·808	2·811	2·814	2·817	2·819	2·822	2·825	2·827
1·5	2·830	2·833	2·835	2·838	2·840	2·843	2·845	2·848	2·851	2·853
1·6	2·856	2·858	2·861	2·863	2·865	2·868	2·870	2·873	2·875	2·878
1·7	2·880	2·882	2·885	2·887	2·889	2·892	2·894	2·896	2·899	2·901
1·8	2·903	2·905	2·908	2·910	2·912	2·914	2·916	2·919	2·921	2·923
1·9	2·925	2·927	2·929	2·932	2·934	2·936	2·938	2·940	2·942	2·944
2·0	2·946	2·948	2·950	2·952	2·954	2·956	2·958	2·960	2·962	2·964
2·1	2·966	2·968	2·970	2·972	2·974	2·976	2·978	2·980	2·982	2·984
2·2	2·986	2·988	2·990	2·992	2·993	2·995	2·997	2·999	3·001	3·003
2·3	3·005	3·007	3·008	3·010	3·012	3·014	3·015	3·017	3·019	3·021
2·4	3·023	3·024	3·026	3·028	3·030	3·031	3·033	3·035	3·037	3·038
97·5	6·960	6·962	6·963	6·965	6·967	6·969	6·970	6·972	6·974	6·976
97·6	6·977	6·979	6·981	6·983	6·985	6·986	6·988	6·990	6·992	6·993
97·7	6·995	6·997	6·999	7·001	7·003	7·005	7·007	7·008	7·010	7·012
97·8	7·014	7·016	7·018	7·020	7·022	7·024	7·026	7·028	7·030	7·032
97·9	7·034	7·036	7·038	7·040	7·042	7·044	7·046	7·048	7·050	7·052
98·0	7·054	7·056	7·058	7·060	7·062	7·064	7·066	7·068	7·071	7·073
98·1	7·075	7·077	7·079	7·081	7·084	7·086	7·088	7·090	7·092	7·095
98·2	7·097	7·099	7·101	7·104	7·106	7·108	7·111	7·113	7·115	7·118
98·3	7·120	7·122	7·125	7·127	7·130	7·132	7·135	7·137	7·139	7·142
98·4	7·144	7·147	7·149	7·152	7·155	7·157	7·160	7·162	7·165	7·167
98·5	7·170	7·173	7·175	7·178	7·181	7·183	7·186	7·189	7·192	7·194
98·6	7·197	7·200	7·203	7·206	7·209	7·212	7·214	7·217	7·220	7·223
98·7	7·226	7·229	7·232	7·235	7·238	7·241	7·245	7·248	7·251	7·254
98·8	7·257	7·260	7·264	7·267	7·270	7·273	7·277	7·280	7·284	7·287
98·9	7·290	7·294	7·297	7·301	7·304	7·308	7·312	7·315	7·319	7·323
99·0	7·326	7·330	7·334	7·338	7·342	7·346	7·349	7·353	7·357	7·362
99·1	7·366	7·370	7·374	7·378	7·382	7·387	7·391	7·395	7·400	7·404
99·2	7·409	7·414	7·418	7·423	7·428	7·432	7·437	7·442	7·447	7·452
99·3	7·457	7·462	7·468	7·473	7·478	7·484	7·489	7·495	7·501	7·506
99·4	7·512	7·518	7·524	7·530	7·536	7·543	7·549	7·556	7·562	7·569
99·5	7·576	7·583	7·590	7·597	7·605	7·612	7·620	7·628	7·636	7·644
99·6	7·652	7·661	7·669	7·678	7·687	7·697	7·706	7·716	7·727	7·737
99·7	7·748	7·759	7·770	7·782	7·794	7·807	7·820	7·834	7·848	7·863
99·8	7·878	7·894	7·911	7·929	7·948	7·968	7·989	8·011	8·036	8·062
99·9	8·090	8·121	8·156	8·195	8·239	8·291	8·353	8·432	8·540	8·719

Example: $u_{\cdot0025}+5\cdot000 = 2\cdot193$, $u_{\cdot9975}+5\cdot000 = 7\cdot807$.

TABLE IV. FRACTILES OF THE t DISTRIBUTION. $t_{1-P} = -t_P$.

P \\ f	PROBABILITY IN PER CENT									
	60	70	80	90	95	97·5	99	99·5	99·9	99·95
1	·325	·727	1·376	3·078	6·314	12·71	31·82	63·66	318·3	636·6
2	·289	·617	1·061	1·886	2·920	4·303	6·965	9·925	22·33	31·60
3	·277	·584	·978	1·638	2·353	3·182	4·541	5·841	10·22	12·94
4	·271	·569	·941	1·533	2·132	2·776	3·747	4·604	7·173	8·610
5	·267	·559	·920	1·476	2·015	2·571	3·365	4·032	5·893	6·859
6	·265	·553	·906	1·440	1·943	2·447	3·143	3·707	5·208	5·959
7	·263	·549	·896	1·415	1·895	2·365	2·998	3·499	4·785	5·405
8	·262	·546	·889	1·397	1·860	2·306	2·896	3·355	4·501	5·041
9	·261	·543	·883	1·383	1·833	2·262	2·821	3·250	4·297	4·781
10	·260	·542	·879	1·372	1·812	2·228	2·764	3·169	4·144	4·587
11	·260	·540	·876	1·363	1·796	2·201	2·718	3·106	4·025	4·437
12	·259	·539	·873	1·356	1·782	2·179	2·681	3·055	3·930	4·318
13	·259	·538	·870	1·350	1·771	2·160	2·650	3·012	3·852	4·221
14	·258	·537	·868	1·345	1·761	2·145	2·624	2·977	3·787	4·140
15	·258	·536	·866	1·341	1·753	2·131	2·602	2·947	3·733	4·073
16	·258	·535	·865	1·337	1·746	2·120	2·583	2·921	3·686	4·015
17	·257	·534	·863	1·333	1·740	2·110	2·567	2·898	3·646	3·965
18	·257	·534	·862	1·330	1·734	2·101	2·552	2·878	3·611	3·922
19	·257	·533	·861	1·328	1·729	2·093	2·539	2·861	3·579	3·883
20	·257	·533	·860	1·325	1·725	2·086	2·528	2·845	3·552	3·850
21	·257	·532	·859	1·323	1·721	2·080	2·518	2·831	3·527	3·819
22	·256	·532	·858	1·321	1·717	2·074	2·508	2·819	3·505	3·792
23	·256	·532	·858	1·319	1·714	2·069	2·500	2·807	3·485	3·767
24	·256	·531	·857	1·318	1·711	2·064	2·492	2·797	3·467	3·745
25	·256	·531	·856	1·316	1·708	2·060	2·485	2·787	3·450	3·725
26	·256	·531	·856	1·315	1·706	2·056	2·479	2·779	3·435	3·707
27	·256	·531	·855	1·314	1·703	2·052	2·473	2·771	3·421	3·690
28	·256	·530	·855	1·313	1·701	2·048	2·467	2·763	3·408	3·674
29	·256	·530	·854	1·311	1·699	2·045	2·462	2·756	3·396	3·659
30	·256	·530	·854	1·310	1·697	2·042	2·457	2·750	3·385	3·646
40	·255	·529	·851	1·303	1·684	2·021	2·423	2·704	3·307	3·551
50	·255	·528	·849	1·298	1·676	2·009	2·403	2·678	3·262	3·495
60	·254	·527	·848	1·296	1·671	2·000	2·390	2·660	3·232	3·460
80	·254	·527	·846	1·292	1·664	1·990	2·374	2·639	3·195	3·415
100	·254	·526	·845	1·290	1·660	1·984	2·365	2·626	3·174	3·389
200	·254	·525	·843	1·286	1·653	1·972	2·345	2·601	3·131	3·339
500	·253	·525	·842	1·283	1·648	1·965	2·334	2·586	3·106	3·310
∞	·253	·524	·842	1·282	1·645	1·960	2·326	2·576	3·090	3·291
$2(1-P)$	80	60	40	20	10	5	2	1	0·2	0·1

Example: $P\{t < 2·086\} = 97·5\%$ for $f = 20$.

$P\{|t| > t_P\} = 2(1-P)$. $P\{|t| > 2·086\} = 5\%$ for $f = 20$.

The greater part of this table is reproduced from Table III of R. A. FISHER and F. YATES: *Statistical Tables*, Oliver and Boyd, Edinburgh, by permission of the authors and the publishers.

TABLE V. FRACTILES OF THE χ^2 DISTRIBUTION

f \ P	PROBABILITY IN PER CENT									
	0·05	0·1	0·5	1·0	2·5	5·0	10·0	20·0	30·0	40·0
1	$\cdot0^6393$	$\cdot0^5157$	$\cdot0^4393$	$\cdot0^3157$	$\cdot0^3982$	$\cdot0^2393$	·0158	·0642	·148	·275
2	$\cdot0^2100$	$\cdot0^2200$	·0100	·0201	·0506	·103	·211	·446	·713	1·02
3	·0153	·0243	·0717	·115	·216	·352	·584	1·00	1·42	1·87
4	·0639	·0908	·207	·297	·484	·711	1·06	1·65	2·19	2·75
5	·158	·210	·412	·554	·831	1·15	1·61	2·34	3·00	3·66
6	·299	·381	·676	·872	1·24	1·64	2·20	3·07	3·83	4·57
7	·485	·598	·989	1·24	1·69	2·17	2·83	3·82	4·67	5·49
8	·710	·857	1·34	1·65	2·18	2·73	3·49	4·59	5·53	6·42
9	·972	1·15	1·73	2·09	2·70	3·33	4·17	5·38	6·39	7·36
10	1·26	1·48	2·16	2·56	3·25	3·94	4·87	6·18	7·27	8·30
11	1·59	1·83	2·60	3·05	3·82	4·57	5·58	6·99	8·15	9·24
12	1·93	2·21	3·07	3·57	4·40	5·23	6·30	7·81	9·03	10·2
13	2·31	2·62	3·57	4·11	5·01	5·89	7·04	8·63	9·93	11·1
14	2·70	3·04	4·07	4·66	5·63	6·57	7·79	9·47	10·8	12·1
15	3·11	3·48	4·60	5·23	6·26	7·26	8·55	10·3	11·7	13·0
16	3·54	3·94	5·14	5·81	6·91	7·96	9·31	11·2	12·6	14·0
17	3·98	4·42	5·70	6·41	7·56	8·67	10·1	12·0	13·5	14·9
18	4·44	4·90	6·26	7·01	8·23	9·39	10·9	12·9	14·4	15·9
19	4·91	5·41	6·84	7·63	8·91	10·1	11·7	13·7	15·4	16·9
20	5·40	5·92	7·43	8·26	9·59	10·9	12·4	14·6	16·3	17·8
21	5·90	6·45	8·03	8·90	10·3	11·6	13·2	15·4	17·2	18·8
22	6·40	6·98	8·64	9·54	11·0	12·3	14·0	16·3	18·1	19·7
23	6·92	7·53	9·26	10·2	11·7	13·1	14·8	17·2	19·0	20·7
24	7·45	8·08	9·89	10·9	12·4	13·8	15·7	18·1	19·9	21·7
25	7·99	8·65	10·5	11·5	13·1	14·6	16·5	18·9	20·9	22·6
26	8·54	9·22	11·2	12·2	13·8	15·4	17·3	19·8	21·8	23·6
27	9·09	9·80	11·8	12·9	14·6	16·2	18·1	20·7	22·7	24·5
28	9·66	10·4	12·5	13·6	15·3	16·9	18·9	21·6	23·6	25·5
29	10·2	11·0	13·1	14·3	16·0	17·7	19·8	22·5	24·6	26·5
30	10·8	11·6	13·8	15·0	16·8	18·5	20·6	23·4	25·5	27·4
31	11·4	12·2	14·5	15·7	17·5	19·3	21·4	24·3	26·4	28·4
32	12·0	12·8	15·1	16·4	18·3	20·1	22·3	25·1	27·4	29·4
33	12·6	13·4	15·8	17·1	19·0	20·9	23·1	26·0	28·3	30·3
34	13·2	14·1	16·5	17·8	19·8	21·7	24·0	26·9	29·2	31·3
35	13·8	14·7	17·2	18·5	20·6	22·5	24·8	27·8	30·2	32·3
36	14·4	15·3	17·9	19·2	21·3	23·3	25·6	28·7	31·1	33·3
37	15·0	16·0	18·6	20·0	22·1	24·1	26·5	29·6	32·1	34·2
38	15·6	16·6	19·3	20·7	22·9	24·9	27·3	30·5	33·0	35·2
39	16·3	17·3	20·0	21·4	23·7	25·7	28·2	31·4	33·9	36·2
40	16·9	17·9	20·7	22·2	24·4	26·5	29·1	32·3	34·9	37·1
41	17·5	18·6	21·4	22·9	25·2	27·3	29·9	33·3	35·8	38·1
42	18·2	19·2	22·1	23·7	26·0	28·1	30·8	34·2	36·8	39·1
43	18·8	19·9	22·9	24·4	26·8	29·0	31·6	35·1	37·7	40·0
44	19·5	20·6	23·6	25·1	27·6	29·8	32·5	36·0	38·6	41·0
45	20·1	21·3	24·3	25·9	28·4	30·6	33·4	36·9	39·6	42·0
46	20·8	21·9	25·0	26·7	29·2	31·4	34·2	37·8	40·5	43·0
47	21·5	22·6	25·8	27·4	30·0	32·3	35·1	38·7	41·5	43·9
48	22·1	23·3	26·5	28·2	30·8	33·1	35·9	39·6	42·4	44·9
49	22·8	24·0	27·2	28·9	31·6	33·9	36·8	40·5	43·4	45·9
50	23·5	24·7	28·0	29·7	32·4	34·8	37·7	41·4	44·3	46·9

Example: $P\{\chi^2 < 4\cdot40\} = 2\cdot5\%$ for $f = 12$.

Approximate formula: $\chi_P^2 \simeq \frac{1}{2}\left(\sqrt{2f-1} + u_P\right)^2$ for $f > 30$.

This table has been abridged from *A Table of Percentage Points of the χ^2 Distribution*

TABLE V. FRACTILES OF THE χ^2 DISTRIBUTION   41

PROBABILITY IN PER CENT

50·0	60·0	70·0	80·0	90·0	95·0	97·5	99·0	99·5	99·9	99·95	P / f
·455	·708	1·07	1·64	2·71	3·84	5·02	6·63	7·88	10·8	12·1	1
1·39	1·83	2·41	3·22	4·61	5·99	7·38	9·21	10·6	13·8	15·2	2
2·37	2·95	3·67	4·64	6·25	7·81	9·35	11·3	12·8	16·3	17·7	3
3·36	4·04	4·88	5·99	7·78	9·49	11·1	13·3	14·9	18·5	20·0	4
4·35	5·13	6·06	7·29	9·24	11·1	12·8	15·1	16·7	20·5	22·1	5
5·35	6·21	7·23	8·56	10·6	12·6	14·4	16·8	18·5	22·5	24·1	6
6·35	7·28	8·38	9·80	12·0	14·1	16·0	18·5	20·3	24·3	26·0	7
7·34	8·35	9·52	11·0	13·4	15·5	17·5	20·1	22·0	26·1	27·9	8
8·34	9·41	10·7	12·2	14·7	16·9	19·0	21·7	23·6	27·9	29·7	9
9·34	10·5	11·8	13·4	16·0	18·3	20·5	23·2	25·2	29·6	31·4	10
10·3	11·5	12·9	14·6	17·3	19·7	21·9	24·7	26·8	31·3	33·1	11
11·3	12·6	14·0	15·8	18·5	21·0	23·3	26·2	28·3	32·9	34·8	12
12·3	13·6	15·1	17·0	19·8	22·4	24·7	27·7	29·8	34·5	36·5	13
13·3	14·7	16·2	18·2	21·1	23·7	26·1	29·1	31·3	36·1	38·1	14
14·3	15·7	17·3	19·3	22·3	25·0	27·5	30·6	32·8	37·7	39·7	15
15·3	16·8	18·4	20·5	23·5	26·3	28·8	32·0	34·3	39·3	41·3	16
16·3	17·8	19·5	21·6	24·8	27·6	30·2	33·4	35·7	40·8	42·9	17
17·3	18·9	20·6	22·8	26·0	28·9	31·5	34·8	37·2	42·3	44·4	18
18·3	19·9	21·7	23·9	27·2	30·1	32·9	36·2	38·6	43·8	46·0	19
19·3	21·0	22·8	25·0	28·4	31·4	34·2	37·6	40·0	45·3	47·5	20
20·3	22·0	23·9	26·2	29·6	32·7	35·5	38·9	41·4	46·8	49·0	21
21·3	23·0	24·9	27·3	30·8	33·9	36·8	40·3	42·8	48·3	50·5	22
22·3	24·1	26·0	28·4	32·0	35·2	38·1	41·6	44·2	49·7	52·0	23
23·3	25·1	27·1	29·6	33·2	36·4	39·4	43·0	45·6	51·2	53·5	24
24·3	26·1	28·2	30·7	34·4	37·7	40·6	44·3	46·9	52·6	54·9	25
25·3	27·2	29·2	31·8	35·6	38·9	41·9	45·6	48·3	54·1	56·4	26
26·3	28·2	30·3	32·9	36·7	40·1	43·2	47·0	49·6	55·5	57·9	27
27·3	29·2	31·4	34·0	37·9	41·3	44·5	48·3	51·0	56·9	59·3	28
28·3	30·3	32·5	35·1	39·1	42·6	45·7	49·6	52·3	58·3	60·7	29
29·3	31·3	33·5	36·3	40·3	43·8	47·0	50·9	53·7	59·7	62·2	30
30·3	32·3	34·6	37·4	41·4	45·0	48·2	52·2	55·0	61·1	63·6	31
31·3	33·4	35·7	38·5	42·6	46·2	49·5	53·5	56·3	62·5	65·0	32
32·3	34·4	36·7	39·6	43·7	47·4	50·7	54·8	57·6	63·9	66·4	33
33·3	35·4	37·8	40·7	44·9	48·6	52·0	56·1	59·0	65·2	67·8	34
34·3	36·5	38·9	41·8	46·1	49·8	53·2	57·3	60·3	66·6	69·2	35
35·3	37·5	39·9	42·9	47·2	51·0	54·4	58·6	61·6	68·0	70·6	36
36·3	38·5	41·0	44·0	48·4	52·2	55·7	59·9	62·9	69·3	72·0	37
37·3	39·6	42·0	45·1	49·5	53·4	56·9	61·2	64·2	70·7	73·4	38
38·3	40·6	43·1	46·2	50·7	54·6	58·1	62·4	65·5	72·1	74·7	39
39·3	41·6	44·2	47·3	51·8	55·8	59·3	63·7	66·8	73·4	76·1	40
40·3	42·7	45·2	48·4	52·9	56·9	60·6	65·0	68·1	74·7	77·5	41
41·3	43·7	46·3	49·5	54·1	58·1	61·8	66·2	69·3	76·1	78·8	42
42·3	44·7	47·3	50·5	55·2	59·3	63·0	67·5	70·6	77·4	80·2	43
43·3	45·7	48·4	51·6	56·4	60·5	64·2	68·7	71·9	78·7	81·5	44
44·3	46·8	49·5	52·7	57·5	61·7	65·4	70·0	73·2	80·1	82·9	45
45·3	47·8	50·5	53·8	58·6	62·8	66·6	71·2	74·4	81·4	84·2	46
46·3	48·8	51·6	54·9	59·8	64·0	67·8	72·4	75·7	82·7	85·6	47
47·3	49·8	52·6	56·0	60·9	65·2	69·0	73·7	77·0	84·0	86·9	48
48·3	50·9	53·7	57·1	62·0	66·3	70·2	74·9	78·2	85·4	88·2	49
49·3	51·9	54·7	58·2	63·2	67·5	71·4	76·2	79·5	86·7	89·6	50

Example: $P\{\chi^2 < 23\cdot3\} = 97\cdot5\%$ for $f = 12$.

Approximate formula: $\chi^2_P \simeq \frac{1}{2}(\sqrt{2f-1}+u_P)^2$ for $f > 30$.

by A. Hald and S. A. Sinkbæk, Skandinavisk Aktuarietidskrift, 1950, 168–175.

TABLE V. FRACTILES OF THE χ^2 DISTRIBUTION

PROBABILITY IN PER CENT

f \ P	0·05	0·1	0·5	1·0	2·5	5·0	10·0	20·0	30·0	40·0	50·0
51	24·1	25·4	28·7	30·5	33·2	35·6	38·6	42·4	45·3	47·8	50·3
52	24·8	26·1	29·5	31·2	34·0	36·4	39·4	43·3	46·2	48·8	51·3
53	25·5	26·8	30·2	32·0	34·8	37·3	40·3	44·2	47·2	49·8	52·3
54	26·2	27·5	31·0	32·8	35·6	38·1	41·2	45·1	48·1	50·8	53·3
55	26·9	28·2	31·7	33·6	36·4	39·0	42·1	46·0	49·1	51·7	54·3
56	27·6	28·9	32·5	34·3	37·2	39·8	42·9	47·0	50·0	52·7	55·3
57	28·2	29·6	33·2	35·1	38·0	40·6	43·8	47·9	51·0	53·7	56·3
58	28·9	30·3	34·0	35·9	38·8	41·5	44·7	48·8	51·9	54·7	57·3
59	29·6	31·0	34·8	36·7	39·7	42·3	45·6	49·7	52·9	55·6	58·3
60	30·3	31·7	35·5	37·5	40·5	43·2	46·5	50·6	53·8	56·6	59·3
61	31·0	32·5	36·3	38·3	41·3	44·0	47·3	51·6	54·8	57·6	60·3
62	31·7	33·2	37·1	39·1	42·1	44·9	48·2	52·5	55·7	58·6	61·3
63	32·5	33·9	37·8	39·9	43·0	45·7	49·1	53·4	56·7	59·6	62·3
64	33·2	34·6	38·6	40·6	43·8	46·6	50·0	54·3	57·6	60·5	63·3
65	33·9	35·4	39·4	41·4	44·6	47·4	50·9	55·3	58·6	61·5	64·3
66	34·6	36·1	40·2	42·2	45·4	48·3	51·8	56·2	59·5	62·5	65·3
67	35·3	36·8	40·9	43·0	46·3	49·2	52·7	57·1	60·5	63·5	66·3
68	36·0	37·6	41·7	43·8	47·1	50·0	53·5	58·0	61·4	64·4	67·3
69	36·7	38·3	42·5	44·6	47·9	50·9	54·4	59·0	62·4	65·4	68·3
70	37·5	39·0	43·3	45·4	48·8	51·7	55·3	59·9	63·3	66·4	69·3
71	38·2	39·8	44·1	46·2	49·6	52·6	56·2	60·8	64·3	67·4	70·3
72	38·9	40·5	44·8	47·1	50·4	53·5	57·1	61·8	65·3	68·4	71·3
73	39·6	41·3	45·6	47·9	51·3	54·3	58·0	62·7	66·2	69·3	72·3
74	40·4	42·0	46·4	48·7	52·1	55·2	58·9	63·6	67·2	70·3	73·3
75	41·1	42·8	47·2	49·5	52·9	56·1	59·8	64·5	68·1	71·3	74·3
76	41·8	43·5	48·0	50·3	53·8	56·9	60·7	65·5	69·1	72·3	75·3
77	42·6	44·3	48·8	51·1	54·6	57·8	61·6	66·4	70·0	73·2	76·3
78	43·3	45·0	49·6	51·9	55·5	58·7	62·5	67·3	71·0	74·2	77·3
79	44·1	45·8	50·4	52·7	56·3	59·5	63·4	68·3	72·0	75·2	78·3
80	44·8	46·5	51·2	53·5	57·2	60·4	64·3	69·2	72·9	76·2	79·3
81	45·5	47·3	52·0	54·4	58·0	61·3	65·2	70·1	73·9	77·2	80·3
82	46·3	48·0	52·8	55·2	58·8	62·1	66·1	71·1	74·8	78·1	81·3
83	47·0	48·8	53·6	56·0	59·7	63·0	67·0	72·0	75·8	79·1	82·3
84	47·8	49·6	54·4	56·8	60·5	63·9	67·9	72·9	76·8	80·1	83·3
85	48·5	50·3	55·2	57·6	61·4	64·7	68·8	73·9	77·7	81·1	84·3
86	49·3	51·1	56·0	58·5	62·2	65·6	69·7	74·8	78·7	82·1	85·3
87	50·0	51·9	56·8	59·3	63·1	66·5	70·6	75·7	79·6	83·0	86·3
88	50·8	52·6	57·6	60·1	63·9	67·4	71·5	76·7	80·6	84·0	87·3
89	51·5	53·4	58·4	60·9	64·8	68·2	72·4	77·6	81·6	85·0	88·3
90	52·3	54·2	59·2	61·8	65·6	69·1	73·3	78·6	82·5	86·0	89·3
91	53·0	54·9	60·0	62·6	66·5	70·0	74·2	79·5	83·5	87·0	90·3
92	53·8	55·7	60·8	63·4	67·4	70·9	75·1	80·4	84·4	88·0	91·3
93	54·5	56·5	61·6	64·2	68·2	71·8	76·0	81·4	85·4	88·9	92·3
94	55·3	57·2	62·4	65·1	69·1	72·6	76·9	82·3	86·4	89·9	93·3
95	56·1	58·0	63·2	65·9	69·9	73·5	77·8	83·2	87·3	90·9	94·3
96	56·8	58·8	64·1	66·7	70·8	74·4	78·7	84·2	88·3	91·9	95·3
97	57·6	59·6	64·9	67·6	71·6	75·3	79·6	85·1	89·2	92·9	96·3
98	58·4	60·4	65·7	68·4	72·5	76·2	80·5	86·1	90·2	93·8	97·3
99	59·1	61·1	66·5	69·2	73·4	77·0	81·4	87·0	91·2	94·8	98·3
100	59·9	61·9	67·3	70·1	74·2	77·9	82·4	87·9	92·1	95·8	99·3

Example: $P\{\chi^2 < 43.8\} = 2\cdot5\%$ for $f = 64$.

Approximate formula: $\chi^2_P \simeq \frac{1}{2}\left(\sqrt{2f-1} + u_P\right)^2$ for $f > 30$.

TABLE V. FRACTILES OF THE χ^2 DISTRIBUTION 43

PROBABILITY IN PER CENT

60·0	70·0	80·0	90·0	95·0	97·5	99·0	99·5	99·9	99·95	$P \diagup f$
52·9	55·8	59·2	64·3	68·7	72·6	77·4	80·7	88·0	90·9	51
53·9	56·8	60·3	65·4	69·8	73·8	78·6	82·0	89·3	92·2	52
55·0	57·9	61·4	66·5	71·0	75·0	79·8	83·3	90·6	93·5	53
56·0	58·9	62·5	67·7	72·2	76·2	81·1	84·5	91·9	94·8	54
57·0	60·0	63·6	68·8	73·3	77·4	82·3	85·7	93·2	96·2	55
58·0	61·0	64·7	69·9	74·5	78·6	83·5	87·0	94·5	97·5	56
59·1	62·1	65·7	71·0	75·6	79·8	84·7	88·2	95·8	98·8	57
60·1	63·1	66·8	72·2	76·8	80·9	86·0	89·5	97·0	100·1	58
61·1	64·2	67·9	73·3	77·9	82·1	87·2	90·7	98·3	101·4	59
62·1	65·2	69·0	74·4	79·1	83·3	88·4	92·0	99·6	102·7	60
63·2	66·3	70·0	75·5	80·2	84·5	89·6	93·2	100·9	104·0	61
64·2	67·3	71·1	76·6	81·4	85·7	90·8	94·4	102·2	105·3	62
65·2	68·4	72·2	77·7	82·5	86·8	92·0	95·6	103·4	106·6	63
66·2	69·4	73·3	78·9	83·7	88·0	93·2	96·9	104·7	107·9	64
67·2	70·5	74·4	80·0	84·8	89·2	94·4	98·1	106·0	109·2	65
68·3	71·5	75·4	81·1	86·0	90·3	95·6	99·3	107·3	110·5	66
69·3	72·6	76·5	82·2	87·1	91·5	96·8	100·6	108·5	111·7	67
70·3	73·6	77·6	83·3	88·3	92·7	98·0	101·8	109·8	113·0	68
71·3	74·6	78·6	84·4	89·4	93·9	99·2	103·0	111·1	114·3	69
72·4	75·7	79·7	85·5	90·5	95·0	100·4	104·2	112·3	115·6	70
73·4	76·7	80·8	86·6	91·7	96·2	101·6	105·4	113·6	116·9	71
74·4	77·8	81·9	87·7	92·8	97·4	102·8	106·6	114·8	118·1	72
75·4	78·8	82·9	88·8	93·9	98·5	104·0	107·9	116·1	119·4	73
76·4	79·9	84·0	90·0	95·1	99·7	105·2	109·1	117·3	120·7	74
77·5	80·9	85·1	91·1	96·2	100·8	106·4	110·3	118·6	121·9	75
78·5	82·0	86·1	92·2	97·4	102·0	107·6	111·5	119·9	123·2	76
79·5	83·0	87·2	93·3	98·5	103·2	108·8	112·7	121·1	124·5	77
80·5	84·0	88·3	94·4	99·6	104·3	110·0	113·9	122·3	125·7	78
81·5	85·1	89·3	95·5	100·7	105·5	111·1	115·1	123·6	127·0	79
82·6	86·1	90·4	96·6	101·9	106·6	112·3	116·3	124·8	128·3	80
83·6	87·2	91·5	97·7	103·0	107·8	113·5	117·5	126·1	129·5	81
84·6	88·2	92·5	98·8	104·1	108·9	114·7	118·7	127·3	130·8	82
85·6	89·2	93·6	99·9	105·3	110·1	115·9	119·9	128·6	132·0	83
86·6	90·3	94·7	101·0	106·4	111·2	117·1	121·1	129·8	133·3	84
87·7	91·3	95·7	102·1	107·5	112·4	118·2	122·3	131·0	134·5	85
88·7	92·4	96·8	103·2	108·6	113·5	119·4	123·5	132·3	135·8	86
89·7	93·4	97·9	104·3	109·8	114·7	120·6	124·7	133·5	137·0	87
90·7	94·4	98·9	105·4	110·9	115·8	121·8	125·9	134·7	138·3	88
91·7	95·5	100·0	106·5	112·0	117·0	122·9	127·1	136·0	139·5	89
92·8	96·5	101·1	107·6	113·1	118·1	124·1	128·3	137·2	140·8	90
93·8	97·6	102·1	108·7	114·3	119·3	125·3	129·5	138·4	142·0	91
94·8	98·6	103·2	109·8	115·4	120·4	126·5	130·7	139·7	143·3	92
95·8	99·6	104·2	110·9	116·5	121·6	127·6	131·9	140·9	144·5	93
96·8	100·7	105·3	111·9	117·6	122·7	128·8	133·1	142·1	145·8	94
97·9	101·7	106·4	113·0	118·8	123·9	130·0	134·2	143·3	147·0	95
98·9	102·8	107·4	114·1	119·9	125·0	131·1	135·4	144·6	148·2	96
99·9	103·8	108·5	115·2	121·0	126·1	132·3	136·6	145·8	149·5	97
100·9	104·8	109·5	116·3	122·1	127·3	133·5	137·8	147·0	150·7	98
101·9	105·9	110·6	117·4	123·2	128·4	134·6	139·0	148·2	151·9	99
102·9	106·9	111·7	118·5	124·3	129·6	135·8	140·2	149·4	153·2	100

Example: $P\{\chi^2 < 88\cdot0\} = 97\cdot5\%$ for $f = 64$.

Approximate formula: $\chi_P^2 \simeq \tfrac{1}{2}(\sqrt{2f-1}+u_P)^2$ for $f > 30$.

TABLE VI. FRACTILES OF THE χ^2/f DISTRIBUTION. $\chi^2/f = s^2/\sigma^2$.

f \ P	PROBABILITY IN PER CENT						PROBABILITY IN PER CENT					
	0.05	0.1	0.5	1.0	2.5	5.0	95.0	97.5	99.0	99.5	99.9	99.95
1	·0000	·0000	·0000	·0002	·0010	·0039	3·8410	5·0240	6·6350	7·8790	10·8280	12·1160
2	·0005	·0010	·0050	·0100	·0253	·0515	2·9955	3·6890	4·6050	5·2985	6·9080	7·6010
3	·0051	·0081	·0239	·0383	·0720	·1173	2·6050	3·1160	3·7817	4·2793	5·4220	5·9100
4	·0160	·0227	·0518	·0742	·1210	·1778	2·3720	2·7858	3·3192	3·7150	4·6168	4·9995
5	·0316	·0420	·0824	·1108	·1662	·2290	2·2140	2·5664	3·0172	3·3500	4·1030	4·4210
6	·0499	·0635	·1127	·1453	·2062	·2725	2·0987	2·4082	2·8020	3·0913	3·7430	4·0172
7	·0693	·0854	·1413	·1770	·2414	·3096	2·0096	2·2876	2·6393	2·8969	3·4746	3·7169
8	·0888	·1071	·1680	·2058	·2725	·3416	1·9384	2·1919	2·5112	2·7444	3·2656	3·4835
9	·1080	·1281	·1928	·2320	·3000	·3694	1·8799	2·1137	2·4073	2·6210	3·0974	3·2962
10	·1265	·1479	·2156	·2558	·3247	·3940	1·8307	2·0483	2·3209	2·5188	2·9588	3·1419
11	·1443	·1667	·2366	·2775	·3469	·4159	1·7886	1·9927	2·2477	2·4325	2·8422	3·0124
12	·1612	·1845	·2562	·2976	·3670	·4355	1·7522	1·9447	2·1848	2·3583	2·7424	2·9018
13	·1773	·2013	·2742	·3159	·3853	·4532	1·7202	1·9028	2·1298	2·2938	2·6560	2·8060
14	·1926	·2172	·2911	·3329	·4021	·4694	1·6918	1·8656	2·0815	2·2371	2·5802	2·7221
15	·2072	·2322	·3067	·3486	·4175	·4841	1·6664	1·8325	2·0385	2·1867	2·5131	2·6479
16	·2210	·2464	·3214	·3632	·4318	·4976	1·6435	1·8028	2·0000	2·1417	2·4532	2·5818
17	·2341	·2598	·3351	·3769	·4449	·5101	1·6228	1·7759	1·9652	2·1011	2·3994	2·5223
18	·2466	·2725	·3481	·3897	·4573	·5217	1·6038	1·7514	1·9336	2·0642	2·3507	2·4686
19	·2585	·2846	·3602	·4017	·4688	·5325	1·5865	1·7291	1·9048	2·0306	2·3063	2·4196
20	·2699	·2961	·3717	·4130	·4796	·5426	1·5705	1·7085	1·8783	1·9998	2·2658	2·3749
21	·2808	·3070	·3826	·4237	·4897	·5520	1·5558	1·6895	1·8539	1·9715	2·2284	2·3338
22	·2911	·3174	·3929	·4337	·4992	·5608	1·5420	1·6719	1·8313	1·9453	2·1940	2·2960
23	·3010	·3273	·4026	·4433	·5082	·5692	1·5292	1·6555	1·8103	1·9209	2·1621	2·2609
24	·3105	·3369	·4119	·4523	·5167	·5770	1·5173	1·6402	1·7908	1·8982	2·1325	2·2283
25	·3196	·3460	·4208	·4610	·5248	·5844	1·5061	1·6258	1·7726	1·8771	2·1048	2·1979
26	·3284	·3547	·4292	·4692	·5325	·5915	1·4956	1·6124	1·7555	1·8573	2·0789	2·1695
27	·3368	·3631	·4373	·4770	·5397	·5982	1·4857	1·5998	1·7394	1·8387	2·0547	2·1429
28	·3449	·3711	·4450	·4845	·5467	·6046	1·4763	1·5879	1·7242	1·8212	2·0319	2·1179
29	·3527	·3788	·4524	·4916	·5533	·6106	1·4675	1·5766	1·7099	1·8047	2·0104	2·0943
30	·3601	·3863	·4596	·4984	·5597	·6164	1·4591	1·5660	1·6964	1·7891	1·9901	2·0720
31	·3674	·3934	·4664	·5050	·5658	·6220	1·4511	1·5559	1·6836	1·7743	1·9709	2·0510
32	·3743	·4003	·4729	·5113	·5716	·6272	1·4436	1·5462	1·6714	1·7602	1·9527	2·0311
33	·3811	·4070	·4792	·5174	·5772	·6323	1·4364	1·5371	1·6599	1·7469	1·9355	2·0122
34	·3876	·4134	·4853	·5232	·5825	·6372	1·4295	1·5284	1·6489	1·7342	1·9190	1·9942
35	·3939	·4197	·4912	·5288	·5877	·6419	1·4229	1·5201	1·6383	1·7221	1·9034	1·9771
36	·4000	·4257	·4969	·5342	·5927	·6464	1·4166	1·5121	1·6283	1·7106	1·8885	1·9608
37	·4059	·4315	·5023	·5395	·5975	·6507	1·4106	1·5045	1·6187	1·6995	1·8742	1·9452
38	·4117	·4371	·5076	·5445	·6021	·6548	1·4048	1·4972	1·6095	1·6890	1·8606	1·9303
39	·4173	·4426	·5127	·5494	·6065	·6588	1·3993	1·4903	1·6007	1·6789	1·8476	1·9160
40	·4226	·4479	·5177	·5541	·6108	·6627	1·3940	1·4836	1·5923	1·6692	1·8350	1·9024
41	·4279	·4530	·5225	·5587	·6150	·6665	1·3888	1·4771	1·5841	1·6598	1·8230	1·8892
42	·4330	·4580	·5271	·5631	·6190	·6701	1·3839	1·4709	1·5763	1·6509	1·8115	1·8767
43	·4380	·4629	·5316	·5674	·6229	·6736	1·3792	1·4649	1·5688	1·6422	1·8004	1·8646
44	·4428	·4676	·5360	·5715	·6267	·6770	1·3746	1·4591	1·5616	1·6339	1·7898	1·8529
45	·4475	·4722	·5402	·5756	·6304	·6803	1·3701	1·4536	1·5546	1·6259	1·7795	1·8417
46	·4520	·4767	·5444	·5795	·6339	·6835	1·3659	1·4482	1·5478	1·6182	1·7696	1·8309
47	·4565	·4811	·5484	·5833	·6374	·6866	1·3617	1·4430	1·5413	1·6107	1·7600	1·8204
48	·4609	·4853	·5523	·5870	·6407	·6895	1·3577	1·4380	1·5351	1·6035	1·7508	1·8104
49	·4651	·4894	·5561	·5906	·6440	·6924	1·3539	1·4331	1·5290	1·5966	1·7418	1·8006
50	·4692	·4935	·5598	·5941	·6471	·6953	1·3501	1·4284	1·5231	1·5898	1·7332	1·7912

Example: $P\{\chi^2/f < 0.3670\} = 2.5\%$ for $f = 12$. $P\{\chi^2/f < 1.9447\} = 97.5\%$ for $f = 12$.

Approximate formula: $\chi^2_P/f \simeq \dfrac{1}{2f}(\sqrt{2f-1}+u_P)^2$ for $f > 30$.

f \ P	PROBABILITY IN PER CENT						PROBABILITY IN PER CENT					
	0·05	0·1	0·5	1·0	2·5	5·0	95·0	97·5	99·0	99·5	99·9	99·95
51	·4733	·4974	·5634	·5975	·6502	·6980	1·3465	1·4238	1·5174	1·5833	1·7249	1·7821
52	·4772	·5012	·5669	·6009	·6532	·7007	1·3429	1·4194	1·5118	1·5769	1·7168	1·7733
53	·4810	·5050	·5704	·6041	·6562	·7033	1·3395	1·4151	1·5065	1·5708	1·7089	1·7648
54	·4848	·5087	·5737	·6073	·6590	·7059	1·3362	1·4110	1·5013	1·5649	1·7013	1·7565
55	·4885	·5122	·5770	·6104	·6618	·7083	1·3329	1·4069	1·4962	1·5591	1·6939	1·7484
56	·4921	·5157	·5802	·6134	·6645	·7107	1·3298	1·4030	1·4913	1·5535	1·6868	1·7406
57	·4956	·5191	·5833	·6163	·6671	·7131	1·3267	1·3992	1·4865	1·5480	1·6798	1·7331
58	·4990	·5225	·5863	·6192	·6697	·7154	1·3238	1·3954	1·4819	1·5427	1·6731	1·7257
59	·5024	·5258	·5893	·6220	·6722	·7176	1·3209	1·3918	1·4774	1·5375	1·6665	1·7185
60	·5057	·5290	·5922	·6248	·6747	·7198	1·3180	1·3883	1·4730	1·5325	1·6601	1·7116
61	·5089	·5321	·5951	·6274	·6771	·7219	1·3153	1·3849	1·4687	1·5276	1·6539	1·7048
62	·5121	·5352	·5979	·6300	·6795	·7240	1·3126	1·3815	1·4645	1·5229	1·6478	1·6982
63	·5152	·5382	·6006	·6326	·6817	·7260	1·3100	1·3783	1·4605	1·5182	1·6419	1·6918
64	·5182	·5411	·6033	·6351	·6840	·7280	1·3074	1·3751	1·4565	1·5137	1·6362	1·6855
65	·5212	·5440	·6059	·6376	·6862	·7300	1·3049	1·3720	1·4526	1·5093	1·6306	1·6794
66	·5241	·5469	·6085	·6400	·6883	·7319	1·3025	1·3689	1·4489	1·5050	1·6251	1·6735
67	·5270	·5496	·6110	·6424	·6905	·7338	1·3001	1·3660	1·4452	1·5008	1·6198	1·6677
68	·5298	·5524	·6134	·6447	·6925	·7356	1·2978	1·3631	1·4416	1·4967	1·6146	1·6620
69	·5325	·5550	·6159	·6469	·6946	·7374	1·2955	1·3602	1·4381	1·4927	1·6095	1·6565
70	·5352	·5577	·6182	·6492	·6965	·7391	1·2933	1·3575	1·4346	1·4888	1·6045	1·6511
71	·5379	·5602	·6205	·6514	·6985	·7408	1·2911	1·3548	1·4313	1·4850	1·5997	1·6458
72	·5405	·5628	·6228	·6535	·7004	·7425	1·2890	1·3521	1·4280	1·4812	1·5949	1·6407
73	·5431	·5653	·6251	·6556	·7023	·7442	1·2869	1·3495	1·4248	1·4776	1·5903	1·6356
74	·5456	·5677	·6273	·6576	·7041	·7458	1·2849	1·3470	1·4216	1·4740	1·5858	1·6307
75	·5481	·5701	·6294	·6597	·7059	·7474	1·2829	1·3445	1·4186	1·4705	1·5813	1·6259
76	·5505	·5724	·6316	·6617	·7077	·7489	1·2809	1·3421	1·4156	1·4670	1·5770	1·6212
77	·5529	·5748	·6336	·6636	·7094	·7505	1·2790	1·3397	1·4126	1·4637	1·5727	1·6166
78	·5553	·5771	·6357	·6655	·7111	·7520	1·2771	1·3374	1·4097	1·4604	1·5686	1·6120
79	·5576	·5793	·6377	·6674	·7128	·7534	1·2753	1·3351	1·4069	1·4572	1·5645	1·6076
80	·5599	·5815	·6396	·6692	·7144	·7549	1·2735	1·3329	1·4041	1·4540	1·5605	1·6033
81	·5621	·5837	·6416	·6711	·7160	·7563	1·2717	1·3307	1·4014	1·4509	1·5566	1·5990
82	·5643	·5858	·6435	·6729	·7176	·7577	1·2700	1·3285	1·3987	1·4479	1·5527	1·5948
83	·5665	·5879	·6454	·6746	·7192	·7591	1·2683	1·3264	1·3961	1·4449	1·5490	1·5908
84	·5687	·5900	·6472	·6763	·7207	·7604	1·2666	1·3243	1·3935	1·4420	1·5453	1·5868
85	·5708	·5920	·6491	·6780	·7222	·7618	1·2650	1·3223	1·3910	1·4391	1·5417	1·5828
86	·5728	·5940	·6508	·6797	·7237	·7631	1·2633	1·3203	1·3885	1·4363	1·5381	1·5790
87	·5749	·5960	·6526	·6814	·7252	·7643	1·2618	1·3183	1·3861	1·4335	1·5346	1·5752
88	·5769	·5979	·6543	·6830	·7266	·7656	1·2602	1·3164	1·3837	1·4308	1·5312	1·5715
89	·5789	·5998	·6561	·6846	·7280	·7668	1·2587	1·3145	1·3814	1·4282	1·5278	1·5678
90	·5808	·6017	·6577	·6862	·7294	·7681	1·2572	1·3126	1·3791	1·4255	1·5245	1·5643
91	·5828	·6036	·6594	·6877	·7308	·7693	1·2557	1·3108	1·3768	1·4230	1·5213	1·5607
92	·5847	·6054	·6610	·6892	·7321	·7705	1·2542	1·3090	1·3746	1·4204	1·5181	1·5573
93	·5865	·6072	·6626	·6907	·7335	·7716	1·2528	1·3072	1·3724	1·4180	1·5150	1·5539
94	·5884	·6090	·6642	·6922	·7348	·7728	1·2514	1·3055	1·3702	1·4155	1·5119	1·5505
95	·5902	·6108	·6658	·6937	·7361	·7739	1·2500	1·3038	1·3681	1·4131	1·5089	1·5473
96	·5920	·6125	·6673	·6951	·7373	·7750	1·2487	1·3021	1·3661	1·4108	1·5059	1·5440
97	·5938	·6142	·6688	·6965	·7386	·7761	1·2473	1·3004	1·3640	1·4084	1·5030	1·5409
98	·5955	·6159	·6703	·6979	·7398	·7772	1·2460	1·2988	1·3620	1·4062	1·5001	1·5377
99	·5973	·6175	·6718	·6993	·7410	·7782	1·2447	1·2972	1·3600	1·4039	1·4973	1·5347
100	·5990	·6192	·6733	·7007	·7422	·7793	1·2434	1·2956	1·3581	1·4017	1·4945	1·5317

Example: $P\{\chi^2/f < 0\cdot6840\} = 2\cdot5\%$ for $f = 64$. $P\{\chi^2/f < 1\cdot3751\} = 97\cdot5\%$ for $f = 64$.

Approximate formula: $\chi_P^2/f \simeq \dfrac{1}{2f}(\sqrt{2f-1}+u_P)^2$ for $f > 30$.

TABLE VI. FRACTILES OF THE χ^2/f DISTRIBUTION. $\chi^2/f = s^2/\sigma^2$.

f \\ P	PROBABILITY IN PER CENT						PROBABILITY IN PER CENT					
	0·05	0·1	0·5	1·0	2·5	5·0	95·0	97·5	99·0	99·5	99·9	99·95
100	·5990	·6192	·6733	·7007	·7422	·7793	1·2434	1·2956	1·3581	1·4017	1·4945	1·5317
105	·6072	·6271	·6802	·7071	·7480	·7843	1·2373	1·2881	1·3488	1·3911	1·4812	1·5173
110	·6148	·6344	·6868	·7132	·7534	·7890	1·2316	1·2811	1·3401	1·3813	1·4689	1·5040
115	·6221	·6414	·6930	·7190	·7584	·7934	1·2263	1·2746	1·3321	1·3722	1·4575	1·4916
120	·6289	·6480	·6988	·7243	·7632	·7975	1·2214	1·2685	1·3246	1·3637	1·4468	1·4801
125	·6353	·6542	·7042	·7294	·7676	·8014	1·2167	1·2627	1·3175	1·3557	1·4368	1·4692
130	·6414	·6600	·7094	·7342	·7718	·8051	1·2124	1·2574	1·3109	1·3484	1·4275	1·4592
135	·6473	·6656	·7143	·7388	·7757	·8085	1·2083	1·2523	1·3047	1·3413	1·4187	1·4496
140	·6528	·6709	·7190	·7431	·7795	·8119	1·2043	1·2475	1·2988	1·3346	1·4104	1·4406
145	·6581	·6760	·7234	·7472	·7831	·8150	1·2007	1·2430	1·2933	1·3284	1·4026	1·4321
150	·6631	·6808	·7276	·7511	·7865	·8180	1·1972	1·2387	1·2880	1·3224	1·3951	1·4241
155	·6679	·6854	·7316	·7549	·7898	·8208	1·1939	1·2346	1·2830	1·3168	1·3881	1·4166
160	·6725	·6898	·7355	·7584	·7930	·8235	1·1907	1·2308	1·2783	1·3114	1·3813	1·4093
165	·6769	·6939	·7392	·7618	·7959	·8260	1·1877	1·2270	1·2737	1·3063	1·3751	1·4024
170	·6811	·6980	·7427	·7651	·7987	·8285	1·1848	1·2235	1·2694	1·3014	1·3690	1·3958
175	·6852	·7019	·7461	·7682	·8015	·8309	1·1821	1·2201	1·2653	1·2968	1·3632	1·3896
180	·6891	·7056	·7494	·7712	·8041	·8332	1·1795	1·2170	1·2614	1·2924	1·3577	1·3836
185	·6929	·7092	·7525	·7741	·8066	·8353	1·1769	1·2138	1·2576	1·2881	1·3523	1·3779
190	·6964	·7127	·7555	·7768	·8090	·8374	1·1745	1·2109	1·2541	1·2840	1·3472	1·3725
195	·6999	·7160	·7584	·7795	·8114	·8394	1·1722	1·2081	1·2506	1·2801	1·3424	1·3672
200	·7033	·7192	·7612	·7821	·8136	·8414	1·1700	1·2053	1·2473	1·2763	1·3377	1·3622
210	·7097	·7254	·7665	·7870	·8179	·8451	1·1657	1·2001	1·2409	1·2692	1·3288	1·3526
220	·7157	·7311	·7715	·7916	·8219	·8485	1·1618	1·1953	1·2351	1·2626	1·3207	1·3438
230	·7213	·7365	·7762	·7959	·8256	·8517	1·1582	1·1908	1·2297	1·2564	1·3131	1·3356
240	·7266	·7415	·7805	·7999	·8291	·8547	1·1547	1·1867	1·2246	1·2507	1·3060	1·3279
250	·7317	·7463	·7847	·8037	·8324	·8576	1·1515	1·1828	1·2198	1·2453	1·2994	1·3207
260	·7364	·7507	·7886	·8073	·8355	·8602	1·1485	1·1791	1·2153	1·2403	1·2931	1·3140
270	·7408	·7550	·7923	·8107	·8384	·8628	1·1457	1·1756	1·2111	1·2356	1·2872	1·3077
280	·7450	·7590	·7958	·8139	·8412	·8652	1·1430	1·1723	1·2071	1·2312	1·2817	1·3017
290	·7491	·7629	·7991	·8170	·8438	·8674	1·1404	1·1692	1·2033	1·2269	1·2764	1·2961
300	·7529	·7665	·8023	·8199	·8463	·8696	1·1380	1·1663	1·1997	1·2229	1·2714	1·2907
350	·7698	·7826	·8160	·8326	·8573	·8790	1·1275	1·1535	1·1843	1·2055	1·2500	1·2676
400	·7836	·7957	·8272	·8429	·8662	·8866	1·1191	1·1433	1·1718	1·1915	1·2378	1·2491
450	·7951	·8066	·8366	·8515	·8736	·8929	1·1121	1·1349	1·1616	1·1801	1·2187	1·2340
500	·8050	·8160	·8446	·8588	·8799	·8983	1·1063	1·1277	1·1530	1·1704	1·2070	1·2214
550	·8135	·8239	·8515	·8651	·8853	·9029	1·1012	1·1216	1·1456	1·1622	1·1968	1·2105
600	·8208	·8310	·8575	·8706	·8900	·9070	1·0968	1·1163	1·1392	1·1550	1·1880	1·2010
650	·8275	·8373	·8629	·8755	·8942	·9106	1·0929	1·1116	1·1335	1·1487	1·1803	1·1927
700	·8334	·8429	·8677	·8799	·8980	·9137	1·0895	1·1074	1·1285	1·1430	1·1734	1·1853
750	·8387	·8480	·8720	·8838	·9013	·9166	1·0864	1·1037	1·1240	1·1380	1·1672	1·1787
800	·8436	·8526	·8759	·8874	·9044	·9192	1·0836	1·1004	1·1200	1·1335	1·1617	1·1728
850	·8480	·8568	·8795	·8906	·9072	·9216	1·0811	1·0973	1·1163	1·1294	1·1567	1·1674
900	·8521	·8606	·8827	·8936	·9097	·9237	1·0788	1·0945	1·1129	1·1256	1·1520	1·1624
950	·8559	·8642	·8858	·8964	·9121	·9257	1·0767	1·0919	1·1098	1·1221	1·1478	1·1579
1000	·8594	·8675	·8886	·8989	·9143	·9276	1·0747	1·0895	1·1070	1·1190	1·1440	1·1538
2000	·8992	·9051	·9204	·9279	·9390	·9486	1·0526	1·0629	1·0750	1·0833	1·1006	1·1074
3000	·9172	·9221	·9348	·9409	·9500	·9579	1·0429	1·0513	1·0611	1·0678	1·0817	1·0872
4000	·9280	·9323	·9433	·9487	·9566	·9635	1·0370	1·0443	1·0527	1·0585	1·0705	1·0752
5000	·9355	·9393	·9493	·9541	·9612	·9673	1·0331	1·0396	1·0471	1·0523	1·0630	1·0671
10000	·9541	·9569	·9640	·9674	·9725	·9769	1·0234	1·0279	1·0332	1·0368	1·0443	1·0472

Example: $P\{\chi^2/f < 0.8324\} = 2.5\%$ for $f = 250$. $P\{\chi^2/f < 1.1828\} = 97.5\%$ for $f = 250$.

Approximate formula: $\chi_P^2/f \simeq \dfrac{1}{2f}(\sqrt{2f-1}+u_P)^2$ for $f > 30$.

TABLE VII₁. 50 PER CENT FRACTILES OF THE v^2 DISTRIBUTION

$$v^2(f_1, f_2) = \frac{\chi_1^2/f_1}{\chi_2^2/f_2} = \frac{s_1^2}{s_2^2}.$$

DEGREES OF FREEDOM FOR THE NUMERATOR (f_1)

	1	2	3	4	5	6	7	8	9	10	15	20	30	50	100	200	500	∞
1	1·00	1·50	1·71	1·82	1·89	1·94	1·98	2·00	2·03	2·04	2·09	2·12	2·15	2·17	2·18	2·19	2·19	2·20
2	·667	1·00	1·13	1·21	1·25	1·28	1·30	1·32	1·33	1·34	1·38	1·39	1·41	1·42	1·43	1·44	1·44	1·44
3	·585	·881	1·00	1·06	1·10	1·13	1·15	1·16	1·17	1·18	1·21	1·23	1·24	1·25	1·26	1·26	1·27	1·27
4	·549	·828	·941	1·00	1·04	1·06	1·08	1·09	1·10	1·11	1·14	1·15	1·16	1·18	1·18	1·19	1·19	1·19
5	·528	·799	·907	·965	1·00	1·02	1·04	1·05	1·06	1·07	1·10	1·11	1·12	1·13	1·14	1·15	1·15	1·15
6	·515	·780	·886	·942	·977	1·00	1·02	1·03	1·04	1·05	1·07	1·08	1·10	1·11	1·11	1·12	1·12	1·12
7	·506	·767	·871	·926	·960	·983	1·00	1·01	1·02	1·03	1·05	1·07	1·08	1·09	1·10	1·10	1·10	1·10
8	·499	·757	·860	·915	·948	·971	·988	1·00	1·01	1·02	1·04	1·05	1·07	1·07	1·08	1·09	1·09	1·09
9	·494	·749	·852	·906	·939	·962	·978	·990	1·00	1·01	1·03	1·04	1·05	1·06	1·07	1·08	1·08	1·08
10	·490	·743	·845	·899	·932	·954	·971	·983	·992	1·00	1·02	1·03	1·05	1·06	1·06	1·07	1·07	1·07
11	·486	·739	·840	·893	·926	·948	·964	·977	·986	·994	1·02	1·03	1·04	1·05	1·06	1·06	1·06	1·06
12	·484	·735	·835	·888	·921	·943	·959	·972	·981	·989	1·01	1·02	1·03	1·04	1·05	1·05	1·06	1·06
13	·481	·731	·832	·885	·917	·939	·955	·967	·977	·984	1·01	1·02	1·03	1·04	1·05	1·05	1·05	1·05
14	·479	·729	·828	·881	·914	·936	·952	·964	·973	·981	1·00	1·01	1·03	1·04	1·04	1·05	1·05	1·05
15	·478	·726	·826	·878	·911	·933	·948	·960	·970	·977	1·00	1·01	1·02	1·03	1·04	1·04	1·04	1·05
16	·476	·724	·823	·876	·908	·930	·946	·958	·967	·975	·997	1·01	1·02	1·03	1·04	1·04	1·04	1·04
17	·475	·722	·821	·874	·906	·928	·943	·955	·965	·972	·995	1·01	1·02	1·03	1·03	1·04	1·04	1·04
18	·474	·721	·819	·872	·904	·926	·941	·953	·962	·970	·992	1·00	1·02	1·02	1·03	1·03	1·04	1·04
19	·473	·719	·818	·870	·902	·924	·939	·951	·961	·968	·990	1·00	1·01	1·02	1·03	1·03	1·03	1·04
20	·472	·718	·816	·868	·900	·922	·938	·950	·959	·966	·989	1·00	1·01	1·02	1·03	1·03	1·03	1·03
22	·470	·715	·814	·866	·898	·919	·935	·947	·956	·963	·986	·997	1·01	1·02	1·02	1·03	1·03	1·03
24	·469	·714	·812	·863	·895	·917	·932	·944	·953	·961	·983	·994	1·01	1·02	1·02	1·02	1·03	1·03
26	·468	·712	·810	·861	·893	·915	·930	·942	·951	·959	·981	·992	1·00	1·01	1·02	1·02	1·02	1·03
28	·467	·711	·808	·860	·892	·913	·929	·940	·950	·957	·979	·990	1·00	1·01	1·02	1·02	1·02	1·02
30	·466	·709	·807	·858	·890	·912	·927	·939	·948	·955	·978	·989	1·00	1·01	1·02	1·02	1·02	1·02
40	·463	·705	·802	·854	·885	·907	·922	·934	·943	·950	·972	·983	·994	1·00	1·01	1·01	1·02	1·02
50	·462	·703	·800	·851	·882	·903	·919	·930	·940	·947	·969	·980	·991	1·00	1·01	1·01	1·01	1·01
60	·461	·701	·798	·849	·880	·901	·917	·928	·937	·945	·967	·978	·989	·998	1·00	1·01	1·01	1·01
80	·459	·699	·795	·846	·878	·899	·914	·926	·935	·942	·964	·975	·986	·995	1·00	1·00	1·01	1·01
100	·458	·698	·794	·845	·876	·897	·913	·924	·933	·940	·962	·973	·984	·993	1·00	1·00	1·01	1·01
200	·457	·696	·791	·842	·873	·894	·910	·921	·930	·937	·959	·970	·981	·990	·997	1·00	1·00	1·00
500	·456	·694	·790	·840	·871	·893	·908	·919	·928	·935	·957	·968	·979	·988	·995	·998	1·00	1·00
∞	·455	·693	·789	·839	·870	·891	·907	·918	·927	·934	·956	·967	·978	·987	·993	·997	·999	1·00

Example: $P\{v^2(8,20) < 0.950\} = 50\%$.

$v^2_{.50}(f_1, f_2) = 1/v^2_{.50}(f_2, f_1)$.   Example: $v^2_{.50}(8,20) = 1/v^2_{.50}(20,8) = 1/1.05 = 0.95$.

Approximate formula for f_1 and f_2 larger than 30: $\log_{10} v^2_{.50}(f_1, f_2) \simeq -0.290 \left(\dfrac{1}{f_1} - \dfrac{1}{f_2}\right)$.

A large part of the tables of fractiles of the v^2-distribution, Tables VII₁–VII₉, has been abridged from *Tables of Percentage Points of the Inverted Beta (F) Distribution*, computed by M. MERRINGTON and C. M. THOMPSON, Biometrika, 33, 1943, 73–88, by permission of the proprietors, or reproduced from Table V of R. A. FISHER and F. YATES: *Statistical Tables*, Oliver and Boyd, Edinburgh, by permission of the authors and the publishers.

TABLE VII₂. 70 PER CENT FRACTILES OF THE v^2 DISTRIBUTION

$$v^2(f_1, f_2) = \frac{\chi_1^2/f_1}{\chi_2^2/f_2} = \frac{s_1^2}{s_2^2}.$$

DEGREES OF FREEDOM FOR THE NUMERATOR (f_1)

f_2	1	2	3	4	5	6	7	8	9	10	15	20	30	50	100	200	500
1	3.85	5.06	5.56	5.83	6.00	6.12	6.20	6.27	6.32	6.36	6.48	6.54	6.61	6.66	6.70	6.72	6.73
2	1.92	2.33	2.47	2.56	2.61	2.64	2.66	2.68	2.69	2.70	2.74	2.75	2.77	2.78	2.79	2.80	2.80
3	1.56	1.85	1.94	1.99	2.01	2.03	2.04	2.05	2.05	2.06	2.08	2.08	2.09	2.10	2.10	2.10	2.11
4	1.42	1.65	1.72	1.75	1.77	1.78	1.79	1.79	1.80	1.80	1.81	1.81	1.82	1.82	1.82	1.82	1.82
5	1.34	1.55	1.60	1.63	1.64	1.65	1.65	1.66	1.66	1.66	1.66	1.66	1.67	1.67	1.67	1.67	1.67
6	1.29	1.48	1.53	1.55	1.56	1.57	1.57	1.57	1.57	1.57	1.57	1.57	1.57	1.57	1.57	1.57	1.57
7	1.25	1.44	1.48	1.50	1.51	1.51	1.51	1.51	1.51	1.51	1.51	1.51	1.51	1.50	1.50	1.50	1.50
8	1.23	1.40	1.45	1.46	1.47	1.47	1.47	1.47	1.47	1.47	1.46	1.46	1.46	1.45	1.45	1.45	1.45
9	1.21	1.38	1.42	1.43	1.44	1.44	1.44	1.43	1.43	1.43	1.43	1.42	1.42	1.42	1.41	1.41	1.41
10	1.20	1.36	1.40	1.41	1.41	1.41	1.41	1.41	1.41	1.41	1.40	1.40	1.39	1.39	1.38	1.38	1.38
11	1.18	1.35	1.38	1.39	1.39	1.39	1.39	1.39	1.39	1.38	1.38	1.37	1.37	1.36	1.36	1.35	1.35
12	1.17	1.33	1.37	1.38	1.38	1.38	1.37	1.37	1.37	1.37	1.36	1.35	1.35	1.34	1.33	1.33	1.33
13	1.17	1.32	1.35	1.36	1.36	1.36	1.36	1.36	1.35	1.35	1.34	1.34	1.33	1.32	1.32	1.31	1.31
14	1.16	1.31	1.35	1.35	1.35	1.35	1.35	1.34	1.34	1.34	1.33	1.32	1.31	1.31	1.30	1.30	1.30
15	1.15	1.31	1.34	1.34	1.34	1.34	1.34	1.33	1.33	1.33	1.32	1.31	1.30	1.29	1.29	1.28	1.28
16	1.15	1.30	1.33	1.33	1.33	1.33	1.33	1.32	1.32	1.32	1.31	1.30	1.29	1.28	1.28	1.27	1.27
17	1.15	1.29	1.32	1.33	1.33	1.32	1.32	1.32	1.31	1.31	1.30	1.29	1.28	1.27	1.26	1.26	1.26
18	1.14	1.29	1.32	1.32	1.32	1.32	1.31	1.31	1.31	1.30	1.29	1.28	1.27	1.26	1.26	1.25	1.25
19	1.14	1.28	1.31	1.32	1.31	1.31	1.31	1.30	1.30	1.30	1.28	1.28	1.27	1.26	1.25	1.24	1.24
20	1.13	1.28	1.31	1.31	1.31	1.31	1.30	1.30	1.29	1.29	1.28	1.27	1.26	1.25	1.24	1.24	1.23
22	1.13	1.27	1.30	1.30	1.30	1.30	1.29	1.29	1.28	1.28	1.27	1.26	1.25	1.24	1.23	1.22	1.22
24	1.12	1.27	1.29	1.29	1.29	1.29	1.28	1.28	1.27	1.27	1.26	1.25	1.24	1.22	1.21	1.21	1.21
26	1.12	1.26	1.29	1.29	1.29	1.28	1.28	1.27	1.27	1.26	1.25	1.24	1.23	1.21	1.20	1.20	1.20
28	1.11	1.26	1.28	1.28	1.28	1.28	1.27	1.27	1.26	1.26	1.24	1.23	1.22	1.21	1.20	1.19	1.19
30	1.11	1.25	1.28	1.28	1.28	1.27	1.27	1.26	1.26	1.25	1.24	1.23	1.21	1.20	1.19	1.18	1.18
40	1.10	1.24	1.26	1.26	1.26	1.25	1.25	1.24	1.24	1.23	1.22	1.20	1.19	1.18	1.16	1.15	1.15
50	1.10	1.23	1.25	1.25	1.25	1.24	1.24	1.23	1.23	1.22	1.20	1.18	1.17	1.15	1.13	1.12	1.12
60	1.09	1.23	1.25	1.25	1.24	1.24	1.23	1.23	1.22	1.22	1.19	1.17	1.15	1.14	1.12	1.11	1.10
80	1.09	1.22	1.24	1.24	1.24	1.23	1.22	1.22	1.21	1.21	1.19	1.17	1.15	1.14	1.12	1.11	1.10
100	1.08	1.22	1.24	1.24	1.23	1.22	1.22	1.21	1.21	1.20	1.18	1.17	1.15	1.13	1.11	1.10	1.09
200	1.08	1.21	1.23	1.23	1.22	1.21	1.21	1.20	1.20	1.19	1.17	1.15	1.13	1.11	1.09	1.08	1.07
500	1.08	1.21	1.23	1.22	1.22	1.21	1.20	1.20	1.19	1.18	1.16	1.14	1.12	1.10	1.08	1.06	1.05
∞	1.07	1.20	1.22	1.22	1.21	1.21	1.20	1.19	1.18	1.18	1.15	1.14	1.12	1.09	1.07	1.05	1.03

DEGREES OF FREEDOM FOR THE DENOMINATOR (f_2)

Example: $P\{v^2(8,20) < 1.30\} = 70\%$.

$v^2_{.30}(f_1, f_2) = 1/v^2_{.70}(f_2, f_1)$. Example: $v^2_{.30}(8,20) = 1/v^2_{.70}(20,8) = 1/1.46 = 0.685$.

Approximate formula for f_1 and f_2 larger than 30: $\log_{10} v^2_{.70}(f_1, f_2) \simeq \dfrac{0.4555}{\sqrt{h-0.55}} - 0.329\left(\dfrac{1}{f_1} - \dfrac{1}{f_2}\right)$, where $\dfrac{1}{h} = \dfrac{1}{2}\left(\dfrac{1}{f_1}\right.$

TABLE VII$_3$. 90 PER CENT FRACTILES OF THE v^2 DISTRIBUTION

$$v^2(f_1, f_2) = \frac{\chi_1^2/f_1}{\chi_2^2/f_2} = \frac{s_1^2}{s_2^2}.$$

DEGREES OF FREEDOM FOR THE NUMERATOR (f_1)

	1	2	3	4	5	6	7	8	9	10	15	20	30	50	100	200	500	∞
1	39·9	49·5	53·6	55·8	57·2	58·2	58·9	59·4	59·9	60·2	61·2	61·7	62·3	62·7	63·0	63·2	63·3	63·3
2	8·53	9·00	9·16	9·24	9·29	9·33	9·35	9·37	9·38	9·39	9·42	9·44	9·46	9·47	9·48	9·49	9·49	9·49
3	5·54	5·46	5·39	5·34	5·31	5·28	5·27	5·25	5·24	5·23	5·20	5·18	5·17	5·15	5·14	5·14	5·14	5·13
4	4·54	4·32	4·19	4·11	4·05	4·01	3·98	3·95	3·94	3·92	3·87	3·84	3·82	3·80	3·78	3·77	3·76	3·76
5	4·06	3·78	3·62	3·52	3·45	3·40	3·37	3·34	3·32	3·30	3·24	3·21	3·17	3·15	3·13	3·12	3·11	3·10
6	3·78	3·46	3·29	3·18	3·11	3·05	3·01	2·98	2·96	2·94	2·87	2·84	2·80	2·77	2·75	2·73	2·73	2·72
7	3·59	3·26	3·07	2·96	2·88	2·83	2·78	2·75	2·72	2·70	2·63	2·59	2·56	2·52	2·50	2·48	2·48	2·47
8	3·46	3·11	2·92	2·81	2·73	2·67	2·62	2·59	2·56	2·54	2·46	2·42	2·38	2·35	2·32	2·31	2·30	2·29
9	3·36	3·01	2·81	2·69	2·61	2·55	2·51	2·47	2·44	2·42	2·34	2·30	2·25	2·22	2·19	2·17	2·17	2·16
10	3·28	2·92	2·73	2·61	2·52	2·46	2·41	2·38	2·35	2·32	2·24	2·20	2·16	2·12	2·09	2·07	2·06	2·06
11	3·23	2·86	2·66	2·54	2·45	2·39	2·34	2·30	2·27	2·25	2·17	2·12	2·08	2·04	2·00	1·99	1·98	1·97
12	3·18	2·81	2·61	2·48	2·39	2·33	2·28	2·24	2·21	2·19	2·10	2·06	2·01	1·97	1·94	1·92	1·91	1·90
13	3·14	2·76	2·56	2·43	2·35	2·28	2·23	2·20	2·16	2·14	2·05	2·01	1·96	1·92	1·88	1·86	1·85	1·85
14	3·10	2·73	2·52	2·39	2·31	2·24	2·19	2·15	2·12	2·10	2·01	1·96	1·91	1·87	1·83	1·82	1·80	1·80
15	3·07	2·70	2·49	2·36	2·27	2·21	2·16	2·12	2·09	2·06	1·97	1·92	1·87	1·83	1·79	1·77	1·76	1·76
16	3·05	2·67	2·46	2·33	2·24	2·18	2·13	2·09	2·06	2·03	1·94	1·89	1·84	1·79	1·76	1·74	1·73	1·72
17	3·03	2·64	2·44	2·31	2·22	2·15	2·10	2·06	2·03	2·00	1·91	1·86	1·81	1·76	1·73	1·71	1·69	1·69
18	3·01	2·62	2·42	2·29	2·20	2·13	2·08	2·04	2·00	1·98	1·89	1·84	1·78	1·74	1·70	1·68	1·67	1·66
19	2·99	2·61	2·40	2·27	2·18	2·11	2·06	2·02	1·98	1·96	1·86	1·81	1·76	1·71	1·67	1·65	1·64	1·63
20	2·97	2·59	2·38	2·25	2·16	2·09	2·04	2·00	1·96	1·94	1·84	1·79	1·74	1·69	1·65	1·63	1·62	1·61
22	2·95	2·56	2·35	2·22	2·13	2·06	2·01	1·97	1·93	1·90	1·81	1·76	1·70	1·65	1·61	1·59	1·58	1·57
24	2·93	2·54	2·33	2·19	2·10	2·04	1·98	1·94	1·91	1·88	1·78	1·73	1·67	1·62	1·58	1·56	1·54	1·53
26	2·91	2·52	2·31	2·17	2·08	2·01	1·96	1·92	1·88	1·86	1·76	1·71	1·65	1·59	1·55	1·53	1·51	1·50
28	2·89	2·50	2·29	2·16	2·06	2·00	1·94	1·90	1·87	1·84	1·74	1·69	1·63	1·57	1·53	1·50	1·49	1·48
30	2·88	2·49	2·28	2·14	2·05	1·98	1·93	1·88	1·85	1·82	1·72	1·67	1·61	1·55	1·51	1·48	1·47	1·46
40	2·84	2·44	2·23	2·09	2·00	1·93	1·87	1·83	1·79	1·76	1·66	1·61	1·54	1·48	1·43	1·41	1·39	1·38
50	2·81	2·41	2·20	2·06	1·97	1·90	1·84	1·80	1·76	1·73	1·63	1·57	1·50	1·44	1·39	1·36	1·34	1·33
60	2·79	2·39	2·18	2·04	1·95	1·87	1·82	1·77	1·74	1·71	1·60	1·54	1·48	1·41	1·36	1·33	1·31	1·29
80	2·77	2·37	2·15	2·02	1·92	1·85	1·79	1·75	1·71	1·68	1·57	1·51	1·44	1·38	1·32	1·28	1·26	1·24
100	2·76	2·36	2·14	2·00	1·91	1·83	1·78	1·73	1·70	1·66	1·56	1·49	1·42	1·35	1·29	1·26	1·23	1·21
200	2·73	2·33	2·11	1·97	1·88	1·80	1·75	1·70	1·66	1·63	1·52	1·46	1·38	1·31	1·24	1·20	1·17	1·14
500	2·72	2·31	2·10	1·96	1·86	1·79	1·73	1·68	1·64	1·61	1·50	1·44	1·36	1·28	1·21	1·16	1·12	1·09
∞	2·71	2·30	2·08	1·94	1·85	1·77	1·72	1·67	1·63	1·60	1·49	1·42	1·34	1·26	1·18	1·13	1·08	1·00

Example: $P\{v^2(8,20) < 2\cdot00\} = 90\%$.

$v^2_{\cdot10}(f_1, f_2) = 1/v^2_{\cdot90}(f_2, f_1)$.   Example: $v^2_{\cdot10}(8,20) = 1/v^2_{\cdot90}(20,8) = 1/2\cdot42 = 0\cdot413$.

pproximate formula for f_1 and f_2 larger than 30: $\log_{10} v^2_{\cdot90}(f_1, f_2) \simeq \dfrac{1\cdot1131}{\sqrt{h-0\cdot77}} - 0\cdot527\left(\dfrac{1}{f_1} - \dfrac{1}{f_2}\right)$, where $\dfrac{1}{h} = \dfrac{1}{2}\left(\dfrac{1}{f_1} + \dfrac{1}{f_2}\right)$.

TABLE VII₁. 95 PER CENT FRACTILES OF THE v^2 DISTRIBUTION. $v^2(f_1, f_2) = \dfrac{\chi_1^2/f_1}{\chi_2^2/f_2} = \dfrac{s_1^2}{s_2^2}$.

DEGREES OF FREEDOM FOR THE NUMERATOR (f_1)

.	1	2	3	4	5	6	7	8	9	10	11	12	13	14	15	16	17
1	161	200	216	225	230	234	237	239	241	242	243	244	245	245	246	246	247
2	18·5	19·0	19·2	19·2	19·3	19·3	19·4	19·4	19·4	19·4	19·4	19·4	19·4	19·4	19·4	19·4	19·4
3	10·1	9·55	9·28	9·12	9·01	8·94	8·89	8·85	8·81	8·79	8·76	8·74	8·73	8·71	8·70	8·69	8·68
4	7·71	6·94	6·59	6·39	6·26	6·16	6·09	6·04	6·00	5·96	5·94	5·91	5·89	5·87	5·86	5·84	5·83
5	6·61	5·79	5·41	5·19	5·05	4·95	4·88	4·82	4·77	4·74	4·70	4·68	4·66	4·64	4·62	4·60	4·59
6	5·99	5·14	4·76	4·53	4·39	4·28	4·21	4·15	4·10	4·06	4·03	4·00	3·98	3·96	3·94	3·92	3·91
7	5·59	4·74	4·35	4·12	3·97	3·87	3·79	3·73	3·68	3·64	3·60	3·57	3·55	3·53	3·51	3·49	3·48
8	5·32	4·46	4·07	3·84	3·69	3·58	3·50	3·44	3·39	3·35	3·31	3·28	3·26	3·24	3·22	3·20	3·19
9	5·12	4·26	3·86	3·63	3·48	3·37	3·29	3·23	3·18	3·14	3·10	3·07	3·05	3·03	3·01	2·99	2·97
10	4·96	4·10	3·71	3·48	3·33	3·22	3·14	3·07	3·02	2·98	2·94	2·91	2·89	2·86	2·85	2·83	2·81
11	4·84	3·98	3·59	3·36	3·20	3·09	3·01	2·95	2·90	2·85	2·82	2·79	2·76	2·74	2·72	2·70	2·69
12	4·75	3·89	3·49	3·26	3·11	3·00	2·91	2·85	2·80	2·75	2·72	2·69	2·66	2·64	2·62	2·60	2·58
13	4·67	3·81	3·41	3·18	3·03	2·92	2·83	2·77	2·71	2·67	2·63	2·60	2·58	2·55	2·53	2·51	2·50
14	4·60	3·74	3·34	3·11	2·96	2·85	2·76	2·70	2·65	2·60	2·57	2·53	2·51	2·48	2·46	2·44	2·43
15	4·54	3·68	3·29	3·06	2·90	2·79	2·71	2·64	2·59	2·54	2·51	2·48	2·45	2·42	2·40	2·38	2·37
16	4·49	3·63	3·24	3·01	2·85	2·74	2·66	2·59	2·54	2·49	2·46	2·42	2·40	2·37	2·35	2·33	2·32
17	4·45	3·59	3·20	2·96	2·81	2·70	2·61	2·55	2·49	2·45	2·41	2·38	2·35	2·33	2·31	2·29	2·27
18	4·41	3·55	3·16	2·93	2·77	2·66	2·58	2·51	2·46	2·41	2·37	2·34	2·31	2·29	2·27	2·25	2·23
19	4·38	3·52	3·13	2·90	2·74	2·63	2·54	2·48	2·42	2·38	2·34	2·31	2·28	2·26	2·23	2·21	2·20
20	4·35	3·49	3·10	2·87	2·71	2·60	2·51	2·45	2·39	2·35	2·31	2·28	2·25	2·22	2·20	2·18	2·17
21	4·32	3·47	3·07	2·84	2·68	2·57	2·49	2·42	2·37	2·32	2·28	2·25	2·22	2·20	2·18	2·16	2·14
22	4·30	3·44	3·05	2·82	2·66	2·55	2·46	2·40	2·34	2·30	2·26	2·23	2·20	2·17	2·15	2·13	2·11
23	4·28	3·42	3·03	2·80	2·64	2·53	2·44	2·37	2·32	2·27	2·23	2·20	2·18	2·15	2·13	2·11	2·09
24	4·26	3·40	3·01	2·78	2·62	2·51	2·42	2·36	2·30	2·25	2·21	2·18	2·15	2·13	2·11	2·09	2·07
25	4·24	3·39	2·99	2·76	2·60	2·49	2·40	2·34	2·28	2·24	2·20	2·16	2·14	2·11	2·09	2·07	2·05
26	4·23	3·37	2·98	2·74	2·59	2·47	2·39	2·32	2·27	2·22	2·18	2·15	2·12	2·09	2·07	2·05	2·03
27	4·21	3·35	2·96	2·73	2·57	2·46	2·37	2·31	2·25	2·20	2·17	2·13	2·10	2·08	2·06	2·04	2·02
28	4·20	3·34	2·95	2·71	2·56	2·45	2·36	2·29	2·24	2·19	2·15	2·12	2·09	2·06	2·04	2·02	2·00
29	4·18	3·33	2·93	2·70	2·55	2·43	2·35	2·28	2·22	2·18	2·14	2·10	2·08	2·05	2·03	2·01	1·99
30	4·17	3·32	2·92	2·69	2·53	2·42	2·33	2·27	2·21	2·16	2·13	2·09	2·06	2·04	2·01	1·99	1·98
32	4·15	3·29	2·90	2·67	2·51	2·40	2·31	2·24	2·19	2·14	2·10	2·07	2·04	2·01	1·99	1·97	1·95
34	4·13	3·28	2·88	2·65	2·49	2·38	2·29	2·23	2·17	2·12	2·08	2·05	2·02	1·99	1·97	1·95	1·93
36	4·11	3·26	2·87	2·63	2·48	2·36	2·28	2·21	2·15	2·11	2·07	2·03	2·00	1·98	1·95	1·93	1·92
38	4·10	3·24	2·85	2·62	2·46	2·35	2·26	2·19	2·14	2·09	2·05	2·02	1·99	1·96	1·94	1·92	1·90
40	4·08	3·23	2·84	2·61	2·45	2·34	2·25	2·18	2·12	2·08	2·04	2·00	1·97	1·95	1·92	1·90	1·89
42	4·07	3·22	2·83	2·59	2·44	2·32	2·24	2·17	2·11	2·06	2·03	1·99	1·96	1·93	1·91	1·89	1·87
44	4·06	3·21	2·82	2·58	2·43	2·31	2·23	2·16	2·10	2·05	2·01	1·98	1·95	1·92	1·90	1·88	1·86
46	4·05	3·20	2·81	2·57	2·42	2·30	2·22	2·15	2·09	2·04	2·00	1·97	1·94	1·91	1·89	1·87	1·85
48	4·04	3·19	2·80	2·57	2·41	2·29	2·21	2·14	2·08	2·03	1·99	1·96	1·93	1·90	1·88	1·86	1·84
50	4·03	3·18	2·79	2·56	2·40	2·29	2·20	2·13	2·07	2·03	1·99	1·95	1·92	1·89	1·87	1·85	1·83
55	4·02	3·16	2·77	2·54	2·38	2·27	2·18	2·11	2·06	2·01	1·97	1·93	1·90	1·88	1·85	1·83	1·81
60	4·00	3·15	2·76	2·53	2·37	2·25	2·17	2·10	2·04	1·99	1·95	1·92	1·89	1·86	1·84	1·82	1·80
65	3·99	3·14	2·75	2·51	2·36	2·24	2·15	2·08	2·03	1·98	1·94	1·90	1·87	1·85	1·82	1·80	1·78
70	3·98	3·13	2·74	2·50	2·35	2·23	2·14	2·07	2·02	1·97	1·93	1·89	1·86	1·84	1·81	1·79	1·77
80	3·96	3·11	2·72	2·49	2·33	2·21	2·13	2·06	2·00	1·95	1·91	1·88	1·84	1·82	1·79	1·77	1·75
90	3·95	3·10	2·71	2·47	2·32	2·20	2·11	2·04	1·99	1·94	1·90	1·86	1·83	1·80	1·78	1·76	1·74
100	3·94	3·09	2·70	2·46	2·31	2·19	2·10	2·03	1·97	1·93	1·89	1·85	1·82	1·79	1·77	1·75	1·73
125	3·92	3·07	2·68	2·44	2·29	2·17	2·08	2·01	1·96	1·91	1·87	1·83	1·80	1·77	1·75	1·72	1·70
150	3·90	3·06	2·66	2·43	2·27	2·16	2·07	2·00	1·94	1·89	1·85	1·82	1·79	1·76	1·73	1·71	1·69
200	3·89	3·04	2·65	2·42	2·26	2·14	2·06	1·98	1·93	1·88	1·84	1·80	1·77	1·74	1·72	1·69	1·67
300	3·87	3·03	2·63	2·40	2·24	2·13	2·04	1·97	1·91	1·86	1·82	1·78	1·75	1·72	1·70	1·68	1·66
500	3·86	3·01	2·62	2·39	2·23	2·12	2·03	1·96	1·90	1·85	1·81	1·77	1·74	1·71	1·69	1·66	1·64
1000	3·85	3·00	2·61	2·38	2·22	2·11	2·02	1·95	1·89	1·84	1·80	1·76	1·73	1·70	1·68	1·65	1·63
∞	3·84	3·00	2·60	2·37	2·21	2·10	2·01	1·94	1·88	1·83	1·79	1·75	1·72	1·69	1·67	1·64	1·62

Example: $P\{v^2(8,20) < 2\cdot45\} = 95\%$.

$v^2_{\cdot05}(f_1, f_2) = 1/v^2_{\cdot95}(f_2, f_1)$.   Example: $v^2_{\cdot05}(8,20) = 1/v^2_{\cdot95}(20,8) = 1/3\cdot15 = 0\cdot317$.

DEGREES OF FREEDOM FOR THE DENOMINATOR (f_2)

TABLE VII$_4$. 95 PER CENT FRACTILES OF THE v^2 DISTRIBUTION. $v^2(f_1, f_2) = \dfrac{\chi_1^2/f_1}{\chi_2^2/f_2} = \dfrac{s_1^2}{s_2^2}$. 51

DEGREES OF FREEDOM FOR THE NUMERATOR (f_1)

20	22	24	26	28	30	35	40	45	50	60	80	100	200	500	∞	f_2
248	249	249	249	250	250	251	251	251	252	252	252	253	254	254	254	1
19·4	19·5	19·5	19·5	19·5	19·5	19·5	19·5	19·5	19·5	19·5	19·5	19·5	19·5	19·5	19·5	2
8·66	8·65	8·64	8·63	8·62	8·62	8·60	8·59	8·59	8·58	8·57	8·56	8·55	8·54	8·53	8·53	3
5·80	5·79	5·77	5·76	5·75	5·75	5·73	5·72	5·71	5·70	5·69	5·67	5·66	5·65	5·64	5·63	4
4·56	4·54	4·53	4·52	4·50	4·50	4·48	4·46	4·45	4·44	4·43	4·41	4·41	4·39	4·37	4·37	5
3·87	3·86	3·84	3·83	3·82	3·81	3·79	3·77	3·76	3·75	3·74	3·72	3·71	3·69	3·68	3·67	6
3·44	3·43	3·41	3·40	3·39	3·38	3·36	3·34	3·33	3·32	3·30	3·29	3·27	3·25	3·24	3·23	7
3·15	3·13	3·12	3·10	3·09	3·08	3·06	3·04	3·03	3·02	3·01	2·99	2·97	2·95	2·94	2·93	8
2·94	2·92	2·90	2·89	2·87	2·86	2·84	2·83	2·81	2·80	2·79	2·77	2·76	2·73	2·72	2·71	9
2·77	2·75	2·74	2·72	2·71	2·70	2·68	2·66	2·65	2·64	2·62	2·60	2·59	2·56	2·55	2·54	10
2·65	2·63	2·61	2·59	2·58	2·57	2·55	2·53	2·52	2·51	2·49	2·47	2·46	2·43	2·42	2·40	11
2·54	2·52	2·51	2·49	2·48	2·47	2·44	2·43	2·41	2·40	2·38	2·36	2·35	2·32	2·31	2·30	12
2·46	2·44	2·42	2·41	2·39	2·38	2·36	2·34	2·33	2·31	2·30	2·27	2·26	2·23	2·22	2·21	13
2·39	2·37	2·35	2·33	2·32	2·31	2·28	2·27	2·25	2·24	2·22	2·20	2·19	2·16	2·14	2·13	14
2·33	2·31	2·29	2·27	2·26	2·25	2·22	2·20	2·19	2·18	2·16	2·14	2·12	2·10	2·08	2·07	15
2·28	2·25	2·24	2·22	2·21	2·19	2·17	2·15	2·14	2·12	2·11	2·08	2·07	2·04	2·02	2·01	16
2·23	2·21	2·19	2·17	2·16	2·15	2·12	2·10	2·09	2·08	2·06	2·03	2·02	1·99	1·97	1·96	17
2·19	2·17	2·15	2·13	2·12	2·11	2·08	2·06	2·05	2·04	2·02	1·99	1·98	1·95	1·93	1·92	18
2·16	2·13	2·11	2·10	2·08	2·07	2·05	2·03	2·01	2·00	1·98	1·96	1·94	1·91	1·89	1·88	19
2·12	2·10	2·08	2·07	2·05	2·04	2·01	1·99	1·98	1·97	1·95	1·92	1·91	1·88	1·86	1·84	20
2·10	2·07	2·05	2·04	2·02	2·01	1·98	1·96	1·95	1·94	1·92	1·89	1·88	1·84	1·82	1·81	21
2·07	2·05	2·03	2·01	2·00	1·98	1·96	1·94	1·92	1·91	1·89	1·86	1·85	1·82	1·80	1·78	22
2·05	2·02	2·00	1·99	1·97	1·96	1·93	1·91	1·90	1·88	1·86	1·84	1·82	1·79	1·77	1·76	23
2·03	2·00	1·98	1·97	1·95	1·94	1·91	1·89	1·88	1·86	1·84	1·82	1·80	1·77	1·75	1·73	24
2·01	1·98	1·96	1·95	1·93	1·92	1·89	1·87	1·86	1·84	1·82	1·80	1·78	1·75	1·73	1·71	25
1·99	1·97	1·95	1·93	1·91	1·90	1·87	1·85	1·84	1·82	1·80	1·78	1·76	1·73	1·71	1·69	26
1·97	1·95	1·93	1·91	1·90	1·88	1·86	1·84	1·82	1·81	1·79	1·76	1·74	1·71	1·69	1·67	27
1·96	1·93	1·91	1·90	1·88	1·87	1·84	1·82	1·80	1·79	1·77	1·74	1·73	1·69	1·67	1·65	28
1·94	1·92	1·90	1·88	1·87	1·85	1·83	1·81	1·79	1·77	1·75	1·73	1·71	1·67	1·65	1·64	29
1·93	1·91	1·89	1·87	1·85	1·84	1·81	1·79	1·77	1·76	1·74	1·71	1·70	1·66	1·64	1·62	30
1·91	1·88	1·86	1·85	1·83	1·82	1·79	1·77	1·75	1·74	1·71	1·69	1·67	1·63	1·61	1·59	32
1·89	1·86	1·84	1·82	1·80	1·80	1·77	1·75	1·73	1·71	1·69	1·66	1·65	1·61	1·59	1·57	34
1·87	1·85	1·82	1·81	1·79	1·78	1·75	1·73	1·71	1·69	1·67	1·64	1·62	1·59	1·56	1·55	36
1·85	1·83	1·81	1·79	1·77	1·76	1·73	1·71	1·69	1·68	1·65	1·62	1·61	1·57	1·54	1·53	38
1·84	1·81	1·79	1·77	1·76	1·74	1·72	1·69	1·67	1·66	1·64	1·61	1·59	1·55	1·53	1·51	40
1·83	1·80	1·78	1·76	1·74	1·73	1·70	1·68	1·66	1·65	1·62	1·59	1·57	1·53	1·51	1·49	42
1·81	1·79	1·77	1·75	1·73	1·72	1·69	1·67	1·65	1·63	1·61	1·58	1·56	1·52	1·49	1·48	44
1·80	1·78	1·76	1·74	1·72	1·71	1·68	1·65	1·64	1·62	1·60	1·57	1·55	1·51	1·48	1·46	46
1·79	1·77	1·75	1·73	1·71	1·70	1·67	1·64	1·62	1·61	1·59	1·56	1·54	1·49	1·47	1·45	48
1·78	1·76	1·74	1·72	1·70	1·69	1·66	1·63	1·61	1·60	1·58	1·54	1·52	1·48	1·46	1·44	50
1·76	1·74	1·72	1·70	1·68	1·67	1·64	1·61	1·59	1·58	1·55	1·52	1·50	1·46	1·43	1·41	55
1·75	1·72	1·70	1·68	1·66	1·65	1·62	1·59	1·57	1·56	1·53	1·50	1·48	1·44	1·41	1·39	60
1·73	1·71	1·69	1·67	1·65	1·63	1·60	1·58	1·56	1·54	1·52	1·49	1·46	1·42	1·39	1·37	65
1·72	1·70	1·67	1·65	1·64	1·62	1·59	1·57	1·55	1·53	1·50	1·47	1·45	1·40	1·37	1·35	70
1·70	1·68	1·65	1·63	1·62	1·60	1·57	1·54	1·52	1·51	1·48	1·45	1·43	1·38	1·35	1·32	80
1·69	1·66	1·64	1·62	1·60	1·59	1·55	1·53	1·51	1·49	1·46	1·43	1·41	1·36	1·32	1·30	90
1·68	1·65	1·63	1·61	1·59	1·57	1·54	1·52	1·49	1·48	1·45	1·41	1·39	1·34	1·31	1·28	100
1·65	1·63	1·60	1·58	1·57	1·55	1·52	1·49	1·47	1·45	1·42	1·39	1·36	1·31	1·27	1·25	125
1·64	1·61	1·59	1·57	1·55	1·53	1·50	1·48	1·45	1·44	1·41	1·37	1·34	1·29	1·25	1·22	150
1·62	1·60	1·57	1·55	1·53	1·52	1·48	1·46	1·43	1·41	1·39	1·35	1·32	1·26	1·22	1·19	200
1·61	1·58	1·55	1·53	1·51	1·50	1·46	1·43	1·41	1·39	1·36	1·32	1·30	1·23	1·19	1·15	300
1·59	1·56	1·54	1·52	1·50	1·48	1·45	1·42	1·40	1·38	1·34	1·30	1·28	1·21	1·16	1·11	500
1·58	1·55	1·53	1·51	1·49	1·47	1·44	1·41	1·38	1·36	1·33	1·29	1·26	1·19	1·13	1·08	1000
1·57	1·54	1·52	1·50	1·48	1·46	1·42	1·39	1·37	1·35	1·32	1·27	1·24	1·17	1·11	1·00	∞

DEGREES OF FREEDOM FOR THE DENOMINATOR (f_2)

Approximate formula for f_1 and f_2 larger than 30: $\log_{10} v^2_{\cdot 95}(f_1, f_2) \simeq \dfrac{1 \cdot 4287}{\sqrt{h - 0 \cdot 95}} - 0 \cdot 681 \left(\dfrac{1}{f_1} - \dfrac{1}{f_2} \right)$, where $\dfrac{1}{h} = \dfrac{1}{2} \left(\dfrac{1}{f_1} + \dfrac{1}{f_2} \right)$.

TABLE VII₅. 97·5 PER CENT FRACTILES OF THE v^2 DISTRIBUTION. $v^2(f_1,f_2) = \dfrac{\chi_1^2/f_1}{\chi_2^2/f_2} = \dfrac{s_1^2}{s_2^2}$.

DEGREES OF FREEDOM FOR THE NUMERATOR (f_1)

Degrees of Freedom for the Denominator (f_2)

f_2	1	2	3	4	5	6	7	8	9	10	11	12	13	14	15	16	17
1	648	800	864	900	922	937	948	957	963	969	973	977	980	983	985	987	989
2	38·5	39·0	39·2	39·2	39·3	39·3	39·4	39·4	39·4	39·4	39·4	39·4	39·4	39·4	39·4	39·4	39·4
3	17·4	16·0	15·4	15·1	14·9	14·7	14·6	14·5	14·5	14·4	14·4	14·3	14·3	14·3	14·3	14·2	14·2
4	12·2	10·6	9·98	9·60	9·36	9·20	9·07	8·98	8·90	8·84	8·79	8·75	8·72	8·69	8·66	8·64	8·62
5	10·0	8·43	7·76	7·39	7·15	6·98	6·85	6·76	6·68	6·62	6·57	6·52	6·49	6·46	6·43	6·41	6·39
6	8·81	7·26	6·60	6·23	5·99	5·82	5·70	5·60	5·52	5·46	5·41	5·37	5·33	5·30	5·27	5·25	5·23
7	8·07	6·54	5·89	5·52	5·29	5·12	4·99	4·90	4·82	4·76	4·71	4·67	4·63	4·60	4·57	4·54	4·52
8	7·57	6·06	5·42	5·05	4·82	4·65	4·53	4·43	4·36	4·30	4·24	4·20	4·16	4·13	4·10	4·08	4·05
9	7·21	5·71	5·08	4·72	4·48	4·32	4·20	4·10	4·03	3·96	3·91	3·87	3·83	3·80	3·77	3·74	3·72
10	6·94	5·46	4·83	4·47	4·24	4·07	3·95	3·85	3·78	3·72	3·66	3·62	3·58	3·55	3·52	3·50	3·47
11	6·72	5·26	4·63	4·28	4·04	3·88	3·76	3·66	3·59	3·53	3·47	3·43	3·39	3·36	3·33	3·30	3·28
12	6·55	5·10	4·47	4·12	3·89	3·73	3·61	3·51	3·44	3·37	3·32	3·28	3·24	3·21	3·18	3·15	3·13
13	6·41	4·97	4·35	4·00	3·77	3·60	3·48	3·39	3·31	3·25	3·20	3·15	3·12	3·08	3·05	3·03	3·00
14	6·30	4·86	4·24	3·89	3·66	3·50	3·38	3·29	3·21	3·15	3·09	3·05	3·01	2·98	2·95	2·92	2·90
15	6·20	4·76	4·15	3·80	3·58	3·41	3·29	3·20	3·12	3·06	3·01	2·96	2·92	2·89	2·86	2·84	2·81
16	6·12	4·69	4·08	3·73	3·50	3·34	3·22	3·12	3·05	2·99	2·93	2·89	2·85	2·82	2·79	2·76	2·74
17	6·04	4·62	4·01	3·66	3·44	3·28	3·16	3·06	2·98	2·92	2·87	2·82	2·79	2·75	2·72	2·70	2·67
18	5·98	4·56	3·95	3·61	3·38	3·22	3·10	3·01	2·93	2·87	2·81	2·77	2·73	2·70	2·67	2·64	2·62
19	5·92	4·51	3·90	3·56	3·33	3·17	3·05	2·96	2·88	2·82	2·76	2·72	2·68	2·65	2·62	2·59	2·57
20	5·87	4·46	3·86	3·51	3·29	3·13	3·01	2·91	2·84	2·77	2·72	2·68	2·64	2·60	2·57	2·55	2·52
21	5·83	4·42	3·82	3·48	3·25	3·09	2·97	2·87	2·80	2·73	2·68	2·64	2·60	2·56	2·53	2·51	2·48
22	5·79	4·38	3·78	3·44	3·22	3·05	2·93	2·84	2·76	2·70	2·65	2·60	2·56	2·53	2·50	2·47	2·45
23	5·75	4·35	3·75	3·41	3·18	3·02	2·90	2·81	2·73	2·67	2·62	2·57	2·53	2·50	2·47	2·44	2·42
24	5·72	4·32	3·72	3·38	3·15	2·99	2·87	2·78	2·70	2·64	2·59	2·54	2·50	2·47	2·44	2·41	2·39
25	5·69	4·29	3·69	3·35	3·13	2·97	2·85	2·75	2·68	2·61	2·56	2·51	2·48	2·44	2·41	2·38	2·36
26	5·66	4·27	3·67	3·33	3·10	2·94	2·82	2·73	2·65	2·59	2·54	2·49	2·45	2·42	2·39	2·36	2·34
27	5·63	4·24	3·65	3·31	3·08	2·92	2·80	2·71	2·63	2·57	2·51	2·47	2·43	2·39	2·36	2·34	2·31
28	5·61	4·22	3·63	3·29	3·06	2·90	2·78	2·69	2·61	2·55	2·49	2·45	2·41	2·37	2·34	2·32	2·29
29	5·59	4·20	3·61	3·27	3·04	2·88	2·76	2·67	2·59	2·53	2·48	2·43	2·39	2·36	2·32	2·30	2·27
30	5·57	4·18	3·59	3·25	3·03	2·87	2·75	2·65	2·57	2·51	2·46	2·41	2·37	2·34	2·31	2·28	2·26
32	5·53	4·15	3·56	3·22	3·00	2·84	2·72	2·62	2·54	2·48	2·43	2·38	2·34	2·31	2·28	2·25	2·22
34	5·50	4·12	3·53	3·19	2·97	2·81	2·69	2·59	2·52	2·45	2·40	2·35	2·31	2·28	2·25	2·22	2·19
36	5·47	4·09	3·51	3·17	2·94	2·79	2·66	2·57	2·49	2·43	2·37	2·33	2·29	2·25	2·22	2·20	2·17
38	5·45	4·07	3·48	3·15	2·92	2·76	2·64	2·55	2·47	2·41	2·35	2·31	2·27	2·23	2·20	2·17	2·15
40	5·42	4·05	3·46	3·13	2·90	2·74	2·62	2·53	2·45	2·39	2·33	2·29	2·25	2·21	2·18	2·15	2·13
42	5·40	4·03	3·45	3·11	2·89	2·73	2·61	2·51	2·44	2·37	2·32	2·27	2·23	2·20	2·16	2·14	2·11
44	5·39	4·02	3·43	3·09	2·87	2·71	2·59	2·50	2·42	2·36	2·30	2·26	2·21	2·18	2·15	2·12	2·10
46	5·37	4·00	3·42	3·08	2·86	2·70	2·58	2·48	2·41	2·34	2·29	2·24	2·20	2·17	2·13	2·11	2·08
48	5·35	3·99	3·40	3·07	2·84	2·69	2·57	2·47	2·39	2·33	2·27	2·23	2·19	2·15	2·12	2·09	2·07
50	5·34	3·98	3·39	3·06	2·83	2·67	2·55	2·46	2·38	2·32	2·26	2·22	2·18	2·14	2·11	2·08	2·06
55	5·31	3·95	3·36	3·03	2·81	2·65	2·53	2·43	2·36	2·29	2·24	2·19	2·15	2·11	2·08	2·05	2·03
60	5·29	3·93	3·34	3·01	2·79	2·63	2·51	2·41	2·33	2·27	2·22	2·17	2·13	2·09	2·06	2·03	2·01
65	5·27	3·91	3·32	2·99	2·77	2·61	2·49	2·39	2·32	2·25	2·20	2·15	2·11	2·07	2·04	2·00	1·99
70	5·25	3·89	3·31	2·98	2·75	2·60	2·48	2·38	2·30	2·24	2·18	2·14	2·10	2·06	2·03	1·97	1·95
80	5·22	3·86	3·28	2·95	2·73	2·57	2·45	2·36	2·28	2·21	2·16	2·11	2·07	2·03	2·00	1·95	1·95
90	5·20	3·84	3·27	2·93	2·71	2·55	2·43	2·34	2·26	2·19	2·14	2·09	2·05	2·02	1·98	1·95	1·93
100	5·18	3·83	3·25	2·92	2·70	2·54	2·42	2·32	2·24	2·18	2·12	2·08	2·04	2·00	1·97	1·94	1·91
125	5·15	3·80	3·22	2·89	2·67	2·51	2·39	2·30	2·22	2·15	2·10	2·05	2·01	1·97	1·94	1·91	1·89
150	5·13	3·78	3·20	2·87	2·65	2·49	2·37	2·28	2·20	2·13	2·08	2·03	1·99	1·95	1·92	1·89	1·87
200	5·10	3·76	3·18	2·85	2·63	2·47	2·35	2·26	2·18	2·11	2·06	2·01	1·97	1·93	1·90	1·87	1·84
300	5·08	3·74	3·16	2·83	2·61	2·45	2·33	2·23	2·16	2·09	2·04	1·99	1·95	1·91	1·88	1·85	1·82
500	5·05	3·72	3·14	2·81	2·59	2·43	2·31	2·22	2·14	2·07	2·02	1·97	1·93	1·89	1·86	1·83	1·80
1000	5·04	3·70	3·13	2·80	2·58	2·42	2·30	2·20	2·13	2·06	2·01	1·96	1·92	1·88	1·85	1·82	1·79
∞	5·02	3·69	3·12	2·79	2·57	2·41	2·29	2·19	2·11	2·05	1·99	1·94	1·90	1·87	1·83	1·80	1·78

Example: $P\{v^2(8,20) < 2\cdot91\} = 97\cdot5\%$.

$v^2_{\cdot025}(f_1,f_2) = 1/v^2_{\cdot975}(f_2,f_1)$.   Example: $v^2_{\cdot025}(8,20) = 1/v^2_{\cdot975}(20,8) = 1/4\cdot00 = 0\cdot250$.

TABLE VII₅. 97·5 PER CENT FRACTILES OF THE v^2 DISTRIBUTION. $v^2(f_1, f_2) = \dfrac{\chi_1^2/f_1}{\chi_2^2/f_2} = \dfrac{s_1^2}{s_2^2}$. 53

DEGREES OF FREEDOM FOR THE NUMERATOR (f_1)

20	22	24	26	28	30	35	40	45	50	60	80	100	200	500	∞	
993	995	997	999	1000	1001	1004	1006	1007	1008	1010	1012	1013	1016	1017	1018	1
39·4	39·5	39·5	39·5	39·5	39·5	39·5	39·5	39·5	39·5	39·5	39·5	39·5	39·5	39·5	39·5	2
14·2	14·1	14·1	14·1	14·1	14·1	14·1	14·0	14·0	14·0	14·0	14·0	14·0	13·9	13·9	13·9	3
8·56	8·53	8·51	8·49	8·48	8·46	8·44	8·41	8·39	8·38	8·36	8·33	8·32	8·29	8·27	8·26	4
6·33	6·30	6·28	6·26	6·24	6·23	6·20	6·18	6·16	6·14	6·12	6·10	6·08	6·05	6·03	6·02	5
5·17	5·14	5·12	5·10	5·08	5·07	5·04	5·01	4·99	4·98	4·96	4·93	4·92	4·88	4·86	4·85	6
4·47	4·44	4·42	4·39	4·38	4·36	4·33	4·31	4·29	4·28	4·25	4·23	4·21	4·18	4·16	4·14	7
4·00	3·97	3·95	3·93	3·91	3·89	3·86	3·84	3·82	3·81	3·78	3·76	3·74	3·70	3·68	3·67	8
3·67	3·64	3·61	3·59	3·58	3·56	3·53	3·51	3·49	3·47	3·45	3·42	3·40	3·37	3·35	3·33	9
3·42	3·39	3·37	3·34	3·33	3·31	3·28	3·26	3·24	3·22	3·20	3·17	3·15	3·12	3·09	3·08	10
3·23	3·20	3·17	3·15	3·13	3·12	3·09	3·06	3·04	3·03	3·00	2·97	2·96	2·92	2·90	2·88	11
3·07	3·04	3·02	3·00	2·98	2·96	2·93	2·91	2·89	2·87	2·85	2·82	2·80	2·76	2·74	2·72	12
2·95	2·92	2·89	2·87	2·85	2·84	2·80	2·78	2·76	2·74	2·72	2·69	2·67	2·63	2·61	2·60	13
2·84	2·81	2·79	2·77	2·75	2·73	2·70	2·67	2·65	2·64	2·61	2·58	2·56	2·53	2·50	2·49	14
2·76	2·73	2·70	2·68	2·66	2·64	2·61	2·58	2·56	2·55	2·52	2·49	2·47	2·44	2·41	2·40	15
2·68	2·65	2·63	2·60	2·58	2·57	2·53	2·51	2·49	2·47	2·45	2·42	2·40	2·36	2·33	2·32	16
2·62	2·59	2·56	2·54	2·52	2·50	2·47	2·44	2·42	2·41	2·38	2·35	2·33	2·29	2·26	2·25	17
2·56	2·53	2·50	2·48	2·46	2·44	2·41	2·38	2·36	2·35	2·32	2·29	2·27	2·23	2·20	2·19	18
2·51	2·48	2·45	2·43	2·41	2·39	2·36	2·33	2·31	2·30	2·27	2·24	2·22	2·18	2·15	2·13	19
2·46	2·43	2·41	2·39	2·37	2·35	2·31	2·29	2·27	2·25	2·22	2·19	2·17	2·13	2·10	2·09	20
2·42	2·39	2·37	2·34	2·33	2·31	2·27	2·25	2·23	2·21	2·18	2·15	2·13	2·09	2·06	2·04	21
2·39	2·36	2·33	2·31	2·29	2·27	2·24	2·21	2·19	2·17	2·14	2·11	2·09	2·05	2·02	2·00	22
2·36	2·33	2·30	2·28	2·26	2·24	2·20	2·18	2·15	2·14	2·11	2·08	2·06	2·01	1·99	1·97	23
2·33	2·30	2·27	2·25	2·23	2·21	2·17	2·15	2·12	2·11	2·08	2·05	2·02	1·98	1·95	1·94	24
2·30	2·27	2·24	2·22	2·20	2·18	2·15	2·12	2·10	2·08	2·05	2·02	2·00	1·95	1·92	1·91	25
2·28	2·24	2·22	2·19	2·17	2·16	2·12	2·09	2·07	2·05	2·03	1·99	1·97	1·92	1·90	1·88	26
2·25	2·22	2·19	2·17	2·15	2·13	2·10	2·07	2·05	2·03	2·00	1·97	1·94	1·90	1·87	1·85	27
2·23	2·20	2·17	2·15	2·13	2·11	2·08	2·05	2·03	2·01	1·98	1·94	1·92	1·88	1·85	1·83	28
2·21	2·18	2·15	2·13	2·11	2·09	2·06	2·03	2·01	1·99	1·96	1·92	1·90	1·86	1·83	1·81	29
2·20	2·16	2·14	2·11	2·09	2·07	2·04	2·01	1·99	1·97	1·94	1·90	1·88	1·84	1·81	1·79	30
2·16	2·13	2·10	2·08	2·06	2·04	2·00	1·98	1·95	1·93	1·91	1·87	1·85	1·80	1·77	1·75	32
2·13	2·10	2·07	2·05	2·03	2·01	1·97	1·95	1·92	1·90	1·88	1·84	1·82	1·77	1·74	1·72	34
2·11	2·08	2·05	2·03	2·00	1·99	1·95	1·92	1·90	1·88	1·85	1·81	1·79	1·74	1·71	1·69	36
2·09	2·05	2·03	2·00	1·98	1·96	1·93	1·90	1·87	1·85	1·82	1·79	1·76	1·71	1·68	1·66	38
2·07	2·03	2·01	1·98	1·96	1·94	1·90	1·88	1·85	1·83	1·80	1·76	1·74	1·69	1·66	1·64	40
2·05	2·02	1·99	1·96	1·94	1·92	1·89	1·86	1·83	1·81	1·78	1·74	1·72	1·67	1·64	1·62	42
2·03	2·00	1·97	1·95	1·93	1·91	1·87	1·84	1·82	1·80	1·77	1·73	1·70	1·65	1·62	1·60	44
2·02	1·99	1·96	1·93	1·91	1·89	1·85	1·82	1·80	1·78	1·75	1·71	1·69	1·63	1·60	1·58	46
2·01	1·97	1·94	1·92	1·90	1·88	1·84	1·81	1·79	1·77	1·73	1·69	1·67	1·62	1·58	1·56	48
1·99	1·96	1·93	1·91	1·88	1·87	1·83	1·80	1·77	1·75	1·72	1·68	1·66	1·60	1·57	1·55	50
1·97	1·93	1·90	1·88	1·86	1·84	1·80	1·77	1·74	1·72	1·69	1·65	1·62	1·57	1·54	1·51	55
1·94	1·91	1·88	1·86	1·83	1·82	1·78	1·74	1·72	1·70	1·67	1·62	1·60	1·54	1·51	1·48	60
1·93	1·89	1·86	1·84	1·82	1·80	1·76	1·72	1·70	1·68	1·65	1·60	1·58	1·52	1·48	1·46	65
1·91	1·88	1·85	1·82	1·80	1·78	1·74	1·71	1·68	1·66	1·63	1·58	1·56	1·50	1·46	1·44	70
1·88	1·85	1·82	1·79	1·77	1·75	1·71	1·68	1·65	1·63	1·60	1·55	1·53	1·47	1·43	1·40	80
1·86	1·83	1·80	1·77	1·75	1·73	1·69	1·66	1·63	1·61	1·58	1·53	1·50	1·44	1·40	1·37	90
1·85	1·81	1·78	1·76	1·74	1·71	1·67	1·64	1·61	1·59	1·56	1·51	1·48	1·42	1·38	1·35	100
1·82	1·79	1·75	1·73	1·71	1·68	1·64	1·61	1·58	1·56	1·52	1·48	1·45	1·38	1·34	1·30	125
1·80	1·77	1·74	1·71	1·69	1·67	1·62	1·59	1·56	1·54	1·50	1·45	1·42	1·35	1·31	1·27	150
1·78	1·74	1·71	1·68	1·66	1·64	1·60	1·56	1·53	1·51	1·47	1·42	1·39	1·32	1·27	1·23	200
1·75	1·72	1·69	1·66	1·64	1·62	1·57	1·54	1·51	1·48	1·45	1·39	1·36	1·28	1·23	1·18	300
1·74	1·70	1·67	1·64	1·62	1·60	1·55	1·51	1·49	1·46	1·42	1·37	1·34	1·25	1·19	1·14	500
1·72	1·69	1·65	1·63	1·60	1·58	1·54	1·50	1·47	1·44	1·41	1·35	1·32	1·23	1·16	1·09	1000
1·71	1·67	1·64	1·61	1·59	1·57	1·52	1·48	1·45	1·43	1·39	1·33	1·30	1·21	1·13	1·00	∞

DEGREES OF FREEDOM FOR THE DENOMINATOR (f_2)

Approximate formula for f_1 and f_2 larger than 30: $\quad \log_{10} v_{\cdot 975}^2 \simeq \dfrac{1 \cdot 7023}{\sqrt{h - 1 \cdot 14}} - 0 \cdot 846 \left(\dfrac{1}{f_1} - \dfrac{1}{f_2} \right)$, where $\dfrac{1}{h} = \dfrac{1}{2} \left(\dfrac{1}{f_1} + \dfrac{1}{f_2} \right)$.

TABLE VII$_6$. 99 PER CENT FRACTILES OF THE v^2 DISTRIBUTION. $v^2(f_1, f_2) = \dfrac{\chi_1^2/f_1}{\chi_2^2/f_2} = \dfrac{s_1^2}{s_2^2}$.

DEGREES OF FREEDOM FOR THE NUMERATOR (f_1)

Multiply the numbers of the first row ($f_2 = 1$) by 10.

f_2	1	2	3	4	5	6	7	8	9	10	11	12	13	14	15	16	17
1	405	500	540	563	576	586	593	598	602	606	608	611	613	614	616	617	618
2	98·5	99·0	99·2	99·2	99·3	99·3	99·4	99·4	99·4	99·4	99·4	99·4	99·4	99·4	99·4	99·4	99·4
3	34·1	30·8	29·5	28·7	28·2	27·9	27·7	27·5	27·3	27·2	27·1	27·1	27·0	26·9	26·9	26·8	26·8
4	21·2	18·0	16·7	16·0	15·5	15·2	15·0	14·8	14·7	14·5	14·4	14·4	14·3	14·2	14·2	14·2	14·1
5	16·3	13·3	12·1	11·4	11·0	10·7	10·5	10·3	10·2	10·1	9·96	9·89	9·82	9·77	9·72	9·68	9·64
6	13·7	10·9	9·78	9·15	8·75	8·47	8·26	8·10	7·98	7·87	7·79	7·72	7·66	7·60	7·56	7·52	7·48
7	12·2	9·55	8·45	7·85	7·46	7·19	6·99	6·84	6·72	6·62	6·54	6·47	6·41	6·36	6·31	6·27	6·24
8	11·3	8·65	7·59	7·01	6·63	6·37	6·18	6·03	5·91	5·81	5·73	5·67	5·61	5·56	5·52	5·48	5·44
9	10·6	8·02	6·99	6·42	6·06	5·80	5·61	5·47	5·35	5·26	5·18	5·11	5·05	5·00	4·96	4·92	4·89
10	10·0	7·56	6·55	5·99	5·64	5·39	5·20	5·06	4·94	4·85	4·77	4·71	4·65	4·60	4·56	4·52	4·49
11	9·65	7·21	6·22	5·67	5·32	5·07	4·89	4·74	4·63	4·54	4·46	4·40	4·34	4·29	4·25	4·21	4·18
12	9·33	6·93	5·95	5·41	5·06	4·82	4·64	4·50	4·39	4·30	4·22	4·16	4·10	4·05	4·01	3·97	3·94
13	9·07	6·70	5·74	5·21	4·86	4·62	4·44	4·30	4·19	4·10	4·02	3·96	3·91	3·86	3·82	3·78	3·75
14	8·86	6·51	5·56	5·04	4·70	4·46	4·28	4·14	4·03	3·94	3·86	3·80	3·75	3·70	3·66	3·62	3·59
15	8·68	6·36	5·42	4·89	4·56	4·32	4·14	4·00	3·89	3·80	3·73	3·67	3·61	3·56	3·52	3·49	3·45
16	8·53	6·23	5·29	4·77	4·44	4·20	4·03	3·89	3·78	3·69	3·62	3·55	3·50	3·45	3·41	3·37	3·34
17	8·40	6·11	5·18	4·67	4·34	4·10	3·93	3·79	3·68	3·59	3·52	3·46	3·40	3·35	3·31	3·27	3·24
18	8·29	6·01	5·09	4·58	4·25	4·01	3·84	3·71	3·60	3·51	3·43	3·37	3·32	3·27	3·23	3·19	3·16
19	8·18	5·93	5·01	4·50	4·17	3·94	3·77	3·63	3·52	3·43	3·36	3·30	3·24	3·19	3·15	3·12	3·08
20	8·10	5·85	4·94	4·43	4·10	3·87	3·70	3·56	3·46	3·37	3·29	3·23	3·18	3·13	3·09	3·05	3·02
21	8·02	5·78	4·87	4·37	4·04	3·81	3·64	3·51	3·40	3·31	3·24	3·17	3·12	3·07	3·03	2·99	2·96
22	7·95	5·72	4·82	4·31	3·99	3·76	3·59	3·45	3·35	3·26	3·18	3·12	3·07	3·02	2·98	2·94	2·91
23	7·88	5·66	4·76	4·26	3·94	3·71	3·54	3·41	3·30	3·21	3·14	3·07	3·02	2·97	2·93	2·89	2·86
24	7·82	5·61	4·72	4·22	3·90	3·67	3·50	3·36	3·26	3·17	3·09	3·03	2·98	2·93	2·89	2·85	2·82
25	7·77	5·57	4·68	4·18	3·86	3·63	3·46	3·32	3·22	3·13	3·06	2·99	2·94	2·89	2·85	2·81	2·78
26	7·72	5·53	4·64	4·14	3·82	3·59	3·42	3·29	3·18	3·09	3·02	2·96	2·90	2·86	2·82	2·78	2·74
27	7·68	5·49	4·60	4·11	3·78	3·56	3·39	3·26	3·15	3·06	2·99	2·93	2·87	2·82	2·78	2·75	2·71
28	7·64	5·45	4·57	4·07	3·75	3·53	3·36	3·23	3·12	3·03	2·96	2·90	2·84	2·79	2·75	2·72	2·68
29	7·60	5·42	4·54	4·04	3·73	3·50	3·33	3·20	3·09	3·00	2·93	2·87	2·81	2·77	2·73	2·69	2·66
30	7·56	5·39	4·51	4·02	3·70	3·47	3·30	3·17	3·07	2·98	2·91	2·84	2·79	2·74	2·70	2·66	2·63
32	7·50	5·34	4·46	3·97	3·65	3·43	3·26	3·13	3·02	2·93	2·86	2·80	2·74	2·70	2·66	2·62	2·58
34	7·44	5·29	4·42	3·93	3·61	3·39	3·22	3·09	2·98	2·89	2·82	2·76	2·70	2·66	2·62	2·58	2·55
36	7·40	5·25	4·38	3·89	3·57	3·35	3·18	3·05	2·95	2·86	2·79	2·72	2·67	2·62	2·58	2·54	2·51
38	7·35	5·21	4·34	3·86	3·54	3·32	3·15	3·02	2·92	2·83	2·75	2·69	2·64	2·59	2·55	2·51	2·48
40	7·31	5·18	4·31	3·83	3·51	3·29	3·12	2·99	2·89	2·80	2·73	2·66	2·61	2·56	2·52	2·48	2·45
42	7·28	5·15	4·29	3·80	3·49	3·27	3·10	2·97	2·86	2·78	2·70	2·64	2·59	2·54	2·50	2·46	2·43
44	7·25	5·12	4·26	3·78	3·47	3·24	3·08	2·95	2·84	2·75	2·68	2·62	2·56	2·52	2·47	2·44	2·40
46	7·22	5·10	4·24	3·76	3·44	3·22	3·06	2·93	2·82	2·73	2·66	2·60	2·54	2·50	2·45	2·42	2·38
48	7·19	5·08	4·22	3·74	3·43	3·20	3·04	2·91	2·80	2·72	2·64	2·58	2·53	2·48	2·44	2·40	2·37
50	7·17	5·06	4·20	3·72	3·41	3·19	3·02	2·89	2·79	2·70	2·63	2·56	2·51	2·46	2·42	2·38	2·35
55	7·12	5·01	4·16	3·68	3·37	3·15	2·98	2·85	2·75	2·66	2·59	2·53	2·47	2·42	2·38	2·34	2·31
60	7·08	4·98	4·13	3·65	3·34	3·12	2·95	2·82	2·72	2·63	2·56	2·50	2·44	2·39	2·35	2·31	2·28
65	7·04	4·95	4·10	3·62	3·31	3·09	2·93	2·80	2·69	2·61	2·53	2·47	2·42	2·37	2·33	2·29	2·26
70	7·01	4·92	4·08	3·60	3·29	3·07	2·91	2·78	2·67	2·59	2·51	2·45	2·40	2·35	2·31	2·27	2·23
80	6·96	4·88	4·04	3·56	3·26	3·04	2·87	2·74	2·64	2·55	2·48	2·42	2·36	2·31	2·27	2·23	2·20
90	6·93	4·85	4·01	3·54	3·23	3·01	2·84	2·72	2·61	2·52	2·45	2·39	2·33	2·29	2·24	2·21	2·17
100	6·90	4·82	3·98	3·51	3·21	2·99	2·82	2·69	2·59	2·50	2·43	2·37	2·31	2·26	2·22	2·19	2·15
125	6·84	4·78	3·94	3·47	3·17	2·95	2·79	2·66	2·55	2·47	2·39	2·33	2·28	2·23	2·19	2·15	2·11
150	6·81	4·75	3·92	3·45	3·14	2·92	2·76	2·63	2·53	2·44	2·37	2·31	2·25	2·20	2·16	2·12	2·09
200	6·76	4·71	3·88	3·41	3·11	2·89	2·73	2·60	2·50	2·41	2·34	2·27	2·22	2·17	2·13	2·09	2·06
300	6·72	4·68	3·85	3·38	3·08	2·86	2·70	2·57	2·47	2·38	2·31	2·24	2·19	2·14	2·10	2·06	2·03
500	6·69	4·65	3·82	3·36	3·05	2·84	2·68	2·55	2·44	2·36	2·28	2·22	2·17	2·12	2·07	2·04	2·00
1000	6·66	4·63	3·80	3·34	3·04	2·82	2·66	2·53	2·43	2·34	2·27	2·20	2·15	2·10	2·06	2·02	1·98
∞	6·63	4·61	3·78	3·32	3·02	2·80	2·64	2·51	2·41	2·32	2·25	2·18	2·13	2·08	2·04	2·00	1·97

DEGREES OF FREEDOM FOR THE DENOMINATOR (f_2)

Example: $P\{v^2(8,20) < 3·56\} = 99\%$.

$v_{.01}^2(f_1, f_2) = 1/v_{.99}^2(f_2, f_1)$.   Example: $v_{.01}^2(8,20) = 1/v_{.99}^2(20,8) = 1/5·36 = 0·187$.

DEGREES OF FREEDOM FOR THE NUMERATOR (f_1)

Multiply the numbers of the first row ($f_2=1$) by 10.

20	22	24	26	28	30	35	40	45	50	60	80	100	200	500	∞	f_2
621	622	623	624	625	626	628	629	630	630	631	633	633	635	636	637	1
99.4	99.5	99.5	99.5	99.5	99.5	99.5	99.5	99.5	99.5	99.5	99.5	99.5	99.5	99.5	99.5	2
26.7	26.6	26.6	26.6	26.5	26.5	26.5	26.4	26.4	26.4	26.3	26.3	26.2	26.2	26.1	26.1	3
14.0	14.0	13.9	13.9	13.9	13.8	13.8	13.7	13.7	13.7	13.7	13.6	13.6	13.5	13.5	13.5	4
9.55	9.51	9.47	9.43	9.40	9.38	9.33	9.29	9.26	9.24	9.20	9.16	9.13	9.08	9.04	9.02	5
7.40	7.35	7.31	7.28	7.25	7.23	7.18	7.14	7.11	7.09	7.06	7.01	6.99	6.93	6.90	6.88	6
6.16	6.11	6.07	6.04	6.02	5.99	5.94	5.91	5.88	5.86	5.82	5.78	5.75	5.70	5.67	5.65	7
5.36	5.32	5.28	5.25	5.22	5.20	5.15	5.12	5.09	5.07	5.03	4.99	4.96	4.91	4.88	4.86	8
4.81	4.77	4.73	4.70	4.67	4.65	4.60	4.57	4.54	4.52	4.48	4.44	4.42	4.36	4.33	4.31	9
4.41	4.36	4.33	4.30	4.27	4.25	4.20	4.17	4.14	4.12	4.08	4.04	4.01	3.96	3.93	3.91	10
4.10	4.06	4.02	3.99	3.96	3.94	3.89	3.86	3.83	3.81	3.78	3.73	3.71	3.66	3.62	3.60	11
3.86	3.82	3.78	3.75	3.72	3.70	3.65	3.62	3.59	3.57	3.54	3.49	3.47	3.41	3.38	3.36	12
3.66	3.62	3.59	3.56	3.53	3.51	3.46	3.43	3.40	3.38	3.34	3.30	3.27	3.22	3.19	3.17	13
3.51	3.46	3.43	3.40	3.37	3.35	3.30	3.27	3.24	3.22	3.18	3.14	3.11	3.06	3.03	3.00	14
3.37	3.33	3.29	3.26	3.24	3.21	3.17	3.13	3.10	3.08	3.05	3.00	2.98	2.92	2.89	2.87	15
3.26	3.22	3.18	3.15	3.12	3.10	3.05	3.02	2.99	2.97	2.93	2.89	2.86	2.81	2.78	2.75	16
3.16	3.12	3.08	3.05	3.03	3.00	2.96	2.92	2.89	2.87	2.83	2.79	2.76	2.71	2.68	2.65	17
3.08	3.03	3.00	2.97	2.94	2.92	2.87	2.84	2.81	2.78	2.75	2.70	2.68	2.62	2.59	2.57	18
3.00	2.96	2.92	2.89	2.87	2.84	2.80	2.76	2.73	2.71	2.67	2.63	2.60	2.55	2.51	2.49	19
2.94	2.90	2.86	2.83	2.80	2.78	2.73	2.69	2.67	2.64	2.61	2.56	2.54	2.48	2.44	2.42	20
2.88	2.84	2.80	2.77	2.74	2.72	2.67	2.64	2.61	2.58	2.55	2.50	2.48	2.42	2.38	2.36	21
2.83	2.78	2.75	2.72	2.69	2.67	2.62	2.58	2.55	2.53	2.50	2.45	2.42	2.36	2.33	2.31	22
2.78	2.74	2.70	2.67	2.64	2.62	2.57	2.54	2.51	2.48	2.45	2.40	2.37	2.32	2.28	2.26	23
2.74	2.70	2.66	2.63	2.60	2.58	2.53	2.49	2.46	2.44	2.40	2.36	2.33	2.27	2.24	2.21	24
2.70	2.66	2.62	2.59	2.56	2.54	2.49	2.45	2.42	2.40	2.36	2.32	2.29	2.23	2.19	2.17	25
2.66	2.62	2.58	2.55	2.53	2.50	2.45	2.42	2.39	2.36	2.33	2.28	2.25	2.19	2.16	2.13	26
2.63	2.59	2.55	2.52	2.49	2.47	2.42	2.38	2.35	2.33	2.29	2.25	2.22	2.16	2.12	2.10	27
2.60	2.56	2.52	2.49	2.46	2.44	2.39	2.35	2.32	2.30	2.26	2.22	2.19	2.13	2.09	2.06	28
2.57	2.53	2.49	2.46	2.44	2.41	2.36	2.33	2.30	2.27	2.23	2.19	2.16	2.10	2.06	2.03	29
2.55	2.51	2.47	2.44	2.41	2.39	2.34	2.30	2.27	2.25	2.21	2.16	2.13	2.07	2.03	2.01	30
2.50	2.46	2.42	2.39	2.36	2.34	2.29	2.25	2.22	2.20	2.16	2.11	2.08	2.02	1.98	1.96	32
2.46	2.42	2.38	2.35	2.32	2.30	2.25	2.21	2.18	2.16	2.12	2.07	2.04	1.98	1.94	1.91	34
2.43	2.38	2.35	2.32	2.29	2.26	2.21	2.17	2.14	2.12	2.08	2.03	2.00	1.94	1.90	1.87	36
2.40	2.35	2.32	2.28	2.26	2.23	2.18	2.14	2.11	2.09	2.05	2.00	1.97	1.90	1.86	1.84	38
2.37	2.33	2.29	2.26	2.23	2.20	2.15	2.11	2.08	2.06	2.02	1.97	1.94	1.87	1.83	1.80	40
2.34	2.30	2.26	2.23	2.20	2.18	2.13	2.09	2.06	2.03	1.99	1.94	1.91	1.85	1.80	1.78	42
2.32	2.28	2.24	2.21	2.18	2.15	2.10	2.06	2.03	2.01	1.97	1.92	1.89	1.82	1.78	1.75	44
2.30	2.26	2.22	2.19	2.16	2.13	2.08	2.04	2.01	1.99	1.95	1.90	1.86	1.80	1.75	1.73	46
2.28	2.24	2.20	2.17	2.14	2.12	2.06	2.02	1.99	1.97	1.93	1.88	1.84	1.78	1.73	1.70	48
2.27	2.22	2.18	2.15	2.12	2.10	2.05	2.01	1.97	1.95	1.91	1.86	1.82	1.76	1.71	1.68	50
2.23	2.18	2.15	2.11	2.08	2.06	2.01	1.97	1.93	1.91	1.87	1.81	1.78	1.71	1.67	1.64	55
2.20	2.15	2.12	2.08	2.05	2.03	1.98	1.94	1.90	1.88	1.84	1.78	1.75	1.68	1.63	1.60	60
2.17	2.13	2.09	2.06	2.03	2.00	1.95	1.91	1.88	1.85	1.81	1.75	1.72	1.65	1.60	1.57	65
2.15	2.11	2.07	2.03	2.01	1.98	1.93	1.89	1.85	1.83	1.78	1.73	1.70	1.62	1.57	1.54	70
2.12	2.07	2.03	2.00	1.97	1.94	1.89	1.85	1.81	1.79	1.75	1.69	1.66	1.58	1.53	1.49	80
2.09	2.04	2.00	1.97	1.94	1.92	1.86	1.82	1.79	1.76	1.72	1.66	1.62	1.54	1.49	1.46	90
2.07	2.02	1.98	1.94	1.92	1.89	1.84	1.80	1.76	1.73	1.69	1.63	1.60	1.52	1.47	1.43	100
2.03	1.98	1.94	1.91	1.88	1.85	1.80	1.76	1.72	1.69	1.65	1.59	1.55	1.47	1.41	1.37	125
2.00	1.96	1.92	1.88	1.85	1.83	1.77	1.73	1.69	1.66	1.62	1.56	1.52	1.43	1.38	1.33	150
1.97	1.93	1.89	1.85	1.82	1.79	1.74	1.69	1.66	1.63	1.58	1.52	1.48	1.39	1.33	1.28	200
1.94	1.89	1.85	1.82	1.79	1.76	1.71	1.66	1.62	1.59	1.55	1.48	1.44	1.35	1.28	1.22	300
1.92	1.87	1.83	1.79	1.76	1.74	1.68	1.63	1.60	1.56	1.52	1.45	1.41	1.31	1.23	1.16	500
1.90	1.85	1.81	1.77	1.74	1.72	1.66	1.61	1.57	1.54	1.50	1.43	1.38	1.28	1.19	1.11	1000
1.88	1.83	1.79	1.76	1.72	1.70	1.64	1.59	1.55	1.52	1.47	1.40	1.36	1.25	1.15	1.00	∞

DEGREES OF FREEDOM FOR THE DENOMINATOR (f_2)

Approximate formula for f_1 and f_2 larger than 30: $\log_{10} v^2_{.99} \simeq \dfrac{2.0206}{\sqrt{h-1.40}} - 1.073\left(\dfrac{1}{f_1} - \dfrac{1}{f_2}\right)$, where $\dfrac{1}{h} = \dfrac{1}{2}\left(\dfrac{1}{f_1} + \dfrac{1}{f_2}\right)$.

DEGREES OF FREEDOM FOR THE NUMERATOR (f_1)

DEGREES OF FREEDOM FOR THE DENOMINATOR (f_2)

Multiply the numbers of the first row ($f_2 = 1$) by 100.

f_2	1	2	3	4	5	6	7	8	9	10	11	12	13	14	15	16	17
1	162	200	216	225	231	234	237	239	241	242	243	244	245	246	246	247	247
2	198	199	199	199	199	199	199	199	199	199	199	199	199	199	199	199	199
3	55·6	49·8	47·5	46·2	45·4	44·8	44·4	44·1	43·9	43·7	43·5	43·4	43·3	43·2	43·1	43·0	42·9
4	31·3	26·3	24·3	23·2	22·5	22·0	21·6	21·4	21·1	21·0	20·8	20·7	20·6	20·5	20·4	20·4	20·3
5	22·8	18·3	16·5	15·6	14·9	14·5	14·2	14·0	13·8	13·6	13·5	13·4	13·3	13·2	13·1	13·1	13·0
6	18·6	14·5	12·9	12·0	11·5	11·1	10·8	10·6	10·4	10·2	10·1	10·0	9·95	9·88	9·81	9·76	9·71
7	16·2	12·4	10·9	10·0	9·52	9·16	8·89	8·68	8·51	8·38	8·27	8·18	8·10	8·03	7·97	7·93	7·87
8	14·7	11·0	9·60	8·81	8·30	7·95	7·69	7·50	7·34	7·21	7·10	7·01	6·94	6·87	6·81	6·76	6·72
9	13·6	10·1	8·72	7·96	7·47	7·13	6·88	6·69	6·54	6·42	6·31	6·23	6·15	6·09	6·03	5·98	5·94
10	12·8	9·43	8·08	7·34	6·87	6·54	6·30	6·12	5·97	5·85	5·75	5·66	5·59	5·53	5·47	5·42	5·38
11	12·2	8·91	7·60	6·88	6·42	6·10	5·86	5·68	5·54	5·42	5·32	5·24	5·16	5·10	5·05	5·00	4·96
12	11·8	8·51	7·23	6·52	6·07	5·76	5·52	5·35	5·20	5·09	4·99	4·91	4·84	4·77	4·72	4·67	4·63
13	11·4	8·19	6·93	6·23	5·79	5·48	5·25	5·08	4·94	4·82	4·72	4·64	4·57	4·51	4·46	4·41	4·37
14	11·1	7·92	6·68	6·00	5·56	5·26	5·03	4·86	4·72	4·60	4·51	4·43	4·36	4·30	4·25	4·20	4·16
15	10·8	7·70	6·48	5·80	5·37	5·07	4·85	4·67	4·54	4·42	4·33	4·25	4·18	4·12	4·07	4·02	3·98
16	10·6	7·51	6·30	5·64	5·21	4·91	4·69	4·52	4·38	4·27	4·18	4·10	4·03	3·97	3·92	3·87	3·83
17	10·4	7·35	6·16	5·50	5·07	4·78	4·56	4·39	4·25	4·14	4·05	3·97	3·90	3·84	3·79	3·75	3·71
18	10·2	7·21	6·03	5·37	4·96	4·66	4·44	4·28	4·14	4·03	3·94	3·86	3·79	3·73	3·68	3·64	3·60
19	10·1	7·09	5·92	5·27	4·85	4·56	4·34	4·18	4·04	3·93	3·84	3·76	3·70	3·64	3·59	3·54	3·50
20	9·94	6·99	5·82	5·17	4·76	4·47	4·26	4·09	3·96	3·85	3·76	3·68	3·61	3·55	3·50	3·46	3·42
21	9·83	6·89	5·73	5·09	4·68	4·39	4·18	4·01	3·88	3·77	3·68	3·60	3·54	3·48	3·43	3·38	3·34
22	9·73	6·81	5·65	5·02	4·61	4·32	4·11	3·94	3·81	3·70	3·61	3·54	3·47	3·41	3·36	3·31	3·27
23	9·63	6·73	5·58	4·95	4·54	4·26	4·05	3·88	3·75	3·64	3·55	3·47	3·41	3·35	3·30	3·25	3·21
24	9·55	6·66	5·52	4·89	4·49	4·20	3·99	3·83	3·69	3·59	3·50	3·42	3·35	3·30	3·25	3·20	3·16
25	9·48	6·60	5·46	4·84	4·43	4·15	3·94	3·78	3·64	3·54	3·45	3·37	3·30	3·25	3·20	3·15	3·11
26	9·41	6·54	5·41	4·79	4·38	4·10	3·89	3·73	3·60	3·49	3·40	3·33	3·26	3·20	3·15	3·11	3·07
27	9·34	6·49	5·36	4·74	4·34	4·06	3·85	3·69	3·56	3·45	3·36	3·28	3·22	3·16	3·11	3·07	3·03
28	9·28	6·44	5·32	4·70	4·30	4·02	3·81	3·65	3·52	3·41	3·32	3·25	3·18	3·12	3·07	3·03	2·99
29	9·23	6·40	5·28	4·66	4·26	3·98	3·77	3·61	3·48	3·38	3·29	3·21	3·15	3·09	3·04	2·99	2·95
30	9·18	6·35	5·24	4·62	4·23	3·95	3·74	3·58	3·45	3·34	3·25	3·18	3·11	3·06	3·01	2·96	2·92
32	9·09	6·28	5·17	4·56	4·17	3·89	3·68	3·52	3·39	3·29	3·20	3·12	3·06	3·00	2·95	2·90	2·86
34	9·01	6·22	5·11	4·50	4·11	3·84	3·63	3·47	3·34	3·24	3·15	3·07	3·01	2·95	2·90	2·85	2·81
36	8·94	6·16	5·06	4·46	4·06	3·79	3·58	3·42	3·30	3·19	3·10	3·03	2·96	2·90	2·85	2·81	2·77
38	8·88	6·11	5·02	4·41	4·02	3·75	3·54	3·39	3·25	3·15	3·06	2·99	2·92	2·87	2·82	2·77	2·73
40	8·83	6·07	4·98	4·37	3·99	3·71	3·51	3·35	3·22	3·12	3·03	2·95	2·89	2·83	2·78	2·74	2·70
42	8·78	6·03	4·94	4·34	3·95	3·68	3·48	3·32	3·19	3·09	3·00	2·92	2·86	2·80	2·75	2·71	2·67
44	8·74	5·99	4·91	4·31	3·92	3·65	3·45	3·29	3·16	3·06	2·97	2·89	2·83	2·77	2·72	2·68	2·64
46	8·70	5·96	4·88	4·28	3·90	3·62	3·42	3·26	3·14	3·03	2·94	2·87	2·80	2·75	2·70	2·65	2·61
48	8·66	5·93	4·85	4·25	3·87	3·60	3·40	3·24	3·11	3·01	2·92	2·85	2·78	2·72	2·67	2·63	2·59
50	8·63	5·90	4·83	4·23	3·85	3·58	3·38	3·22	3·09	2·99	2·90	2·82	2·76	2·70	2·65	2·61	2·57
55	8·55	5·84	4·77	4·18	3·80	3·53	3·33	3·17	3·05	2·94	2·85	2·78	2·71	2·66	2·61	2·56	2·52
60	8·49	5·80	4·73	4·14	3·76	3·49	3·29	3·13	3·01	2·90	2·82	2·74	2·68	2·62	2·57	2·53	2·49
65	8·44	5·75	4·68	4·11	3·73	3·46	3·26	3·10	2·98	2·87	2·79	2·71	2·65	2·59	2·54	2·49	2·45
70	8·40	5·72	4·65	4·08	3·70	3·43	3·23	3·08	2·95	2·85	2·76	2·68	2·62	2·56	2·51	2·47	2·43
80	8·33	5·67	4·61	4·03	3·65	3·39	3·19	3·03	2·91	2·80	2·72	2·64	2·58	2·52	2·47	2·43	2·39
90	8·28	5·62	4·57	3·99	3·62	3·35	3·15	3·00	2·87	2·77	2·68	2·61	2·54	2·49	2·44	2·39	2·35
100	8·24	5·59	4·54	3·96	3·59	3·33	3·13	2·97	2·85	2·74	2·66	2·58	2·52	2·46	2·41	2·37	2·33
125	8·17	5·53	4·49	3·91	3·54	3·28	3·08	2·93	2·80	2·70	2·61	2·54	2·47	2·42	2·37	2·32	2·28
150	8·12	5·49	4·45	3·88	3·51	3·25	3·05	2·89	2·77	2·67	2·58	2·51	2·44	2·38	2·33	2·29	2·25
200	8·06	5·44	4·41	3·84	3·47	3·21	3·01	2·85	2·73	2·63	2·54	2·47	2·40	2·35	2·30	2·25	2·21
300	8·00	5·39	4·37	3·80	3·43	3·17	2·97	2·81	2·69	2·59	2·51	2·43	2·37	2·31	2·26	2·21	2·17
500	7·95	5·36	4·33	3·76	3·40	3·14	2·94	2·79	2·66	2·56	2·48	2·40	2·34	2·28	2·23	2·19	2·14
1000	7·92	5·33	4·31	3·74	3·37	3·11	2·92	2·77	2·64	2·54	2·45	2·38	2·32	2·26	2·21	2·16	2·12
∞	7·88	5·30	4·28	3·72	3·35	3·09	2·90	2·74	2·62	2·52	2·43	2·36	2·29	2·24	2·19	2·14	2·10

Example: $P\{v^2(8,20) < 4·09\} = 99·5\%$.

$v^2_{·005}(f_1,f_2) = 1/v^2_{·995}(f_2,f_1)$.   Example: $v^2_{·005}(8,20) = 1/v^2_{·995}(20,8) = 1/6·61 = 0·151$.

DEGREES OF FREEDOM FOR THE NUMERATOR (f_1)

Multiply the numbers of the first row ($f_2 = 1$) by 100.

20	22	24	26	28	30	35	40	45	50	60	80	100	200	500	∞	f_2
248	249	249	250	250	250	251	251	252	252	253	253	253	254	254	255	1
199	199	199	199	199	199	199	199	199	199	199	199	199	199	200	200	2
42·8	42·7	42·6	42·6	42·5	42·5	42·4	42·3	42·3	42·2	42·1	42·1	42·0	41·9	41·9	41·8	3
20·2	20·1	20·0	20·0	19·9	19·9	19·8	19·8	19·7	19·7	19·6	19·5	19·5	19·4	19·4	19·3	4
12·9	12·8	12·8	12·7	12·7	12·7	12·6	12·5	12·5	12·5	12·4	12·3	12·3	12·2	12·2	12·1	5
9·59	9·53	9·47	9·43	9·39	9·36	9·29	9·24	9·20	9·17	9·12	9·06	9·03	8·95	8·91	8·88	6
7·75	7·69	7·64	7·60	7·57	7·53	7·47	7·42	7·38	7·35	7·31	7·25	7·22	7·15	7·10	7·08	7
6·61	6·55	6·50	6·46	6·43	6·40	6·33	6·29	6·25	6·22	6·18	6·12	6·09	6·02	5·98	5·95	8
5·83	5·78	5·73	5·69	5·65	5·62	5·56	5·52	5·48	5·45	5·41	5·36	5·32	5·26	5·21	5·19	9
5·27	5·22	5·17	5·13	5·10	5·07	5·01	4·97	4·93	4·90	4·86	4·80	4·77	4·71	4·67	4·64	10
4·86	4·80	4·76	4·72	4·68	4·65	4·60	4·55	4·52	4·49	4·44	4·39	4·36	4·29	4·25	4·23	11
4·53	4·48	4·43	4·39	4·36	4·33	4·27	4·23	4·19	4·17	4·12	4·07	4·04	3·97	3·93	3·90	12
4·27	4·22	4·17	4·13	4·10	4·07	4·01	3·97	3·94	3·91	3·87	3·81	3·78	3·71	3·67	3·65	13
4·06	4·01	3·96	3·92	3·89	3·86	3·80	3·76	3·73	3·70	3·66	3·60	3·57	3·50	3·46	3·44	14
3·88	3·83	3·79	3·75	3·72	3·69	3·63	3·58	3·55	3·52	3·48	3·43	3·39	3·33	3·29	3·26	15
3·73	3·68	3·64	3·60	3·57	3·54	3·48	3·44	3·40	3·37	3·33	3·28	3·25	3·18	3·14	3·11	16
3·61	3·56	3·51	3·47	3·44	3·41	3·35	3·31	3·28	3·25	3·21	3·15	3·12	3·05	3·01	2·98	17
3·50	3·45	3·40	3·36	3·33	3·30	3·25	3·20	3·17	3·14	3·10	3·04	3·01	2·94	2·90	2·87	18
3·40	3·35	3·31	3·27	3·24	3·21	3·15	3·11	3·07	3·04	3·00	2·95	2·91	2·85	2·80	2·78	19
3·32	3·27	3·22	3·18	3·15	3·12	3·07	3·02	2·99	2·96	2·92	2·86	2·83	2·76	2·72	2·69	20
3·24	3·19	3·15	3·11	3·08	3·05	2·99	2·95	2·91	2·88	2·84	2·78	2·75	2·68	2·64	2·61	21
3·18	3·12	3·08	3·04	3·01	2·98	2·92	2·88	2·84	2·82	2·77	2·72	2·69	2·62	2·57	2·55	22
3·12	3·06	3·02	2·98	2·95	2·92	2·86	2·82	2·78	2·76	2·71	2·66	2·62	2·56	2·51	2·48	23
3·06	3·01	2·97	2·93	2·90	2·87	2·81	2·77	2·73	2·70	2·66	2·60	2·57	2·50	2·46	2·43	24
3·01	2·96	2·92	2·88	2·85	2·82	2·76	2·72	2·68	2·65	2·61	2·55	2·52	2·45	2·41	2·38	25
2·97	2·92	2·87	2·83	2·80	2·77	2·72	2·67	2·64	2·61	2·56	2·51	2·47	2·40	2·36	2·33	26
2·93	2·88	2·83	2·79	2·76	2·73	2·67	2·63	2·59	2·57	2·52	2·47	2·43	2·36	2·32	2·29	27
2·89	2·84	2·79	2·76	2·72	2·69	2·64	2·59	2·56	2·53	2·48	2·43	2·39	2·32	2·28	2·25	28
2·86	2·80	2·76	2·72	2·69	2·66	2·60	2·56	2·52	2·49	2·45	2·39	2·36	2·28	2·24	2·21	29
2·82	2·77	2·73	2·69	2·66	2·63	2·57	2·52	2·49	2·46	2·42	2·36	2·32	2·25	2·21	2·18	30
2·77	2·71	2·67	2·63	2·60	2·57	2·51	2·47	2·43	2·40	2·36	2·30	2·26	2·19	2·15	2·11	32
2·72	2·66	2·62	2·58	2·55	2·52	2·46	2·42	2·38	2·35	2·30	2·25	2·21	2·14	2·09	2·06	34
2·67	2·62	2·58	2·54	2·50	2·48	2·42	2·37	2·33	2·30	2·26	2·20	2·17	2·09	2·04	2·01	36
2·63	2·58	2·54	2·50	2·47	2·44	2·38	2·33	2·29	2·27	2·22	2·16	2·12	2·05	2·00	1·97	38
2·60	2·55	2·50	2·46	2·43	2·40	2·34	2·30	2·26	2·23	2·18	2·12	2·09	2·01	1·96	1·93	40
2·57	2·52	2·47	2·43	2·40	2·37	2·31	2·26	2·23	2·20	2·15	2·09	2·06	1·98	1·93	1·90	42
2·54	2·49	2·44	2·40	2·37	2·34	2·28	2·24	2·20	2·17	2·12	2·06	2·03	1·95	1·90	1·87	44
2·51	2·46	2·42	2·38	2·34	2·32	2·26	2·21	2·17	2·14	2·10	2·04	2·00	1·92	1·87	1·84	46
2·49	2·44	2·39	2·36	2·32	2·29	2·23	2·19	2·15	2·12	2·07	2·01	1·97	1·89	1·84	1·81	48
2·47	2·42	2·37	2·33	2·30	2·27	2·21	2·16	2·13	2·10	2·05	1·99	1·95	1·87	1·82	1·79	50
2·42	2·37	2·33	2·29	2·26	2·23	2·16	2·12	2·08	2·05	2·00	1·94	1·90	1·82	1·77	1·73	55
2·39	2·33	2·29	2·25	2·22	2·19	2·13	2·08	2·04	2·01	1·96	1·90	1·86	1·78	1·73	1·69	60
2·36	2·30	2·26	2·22	2·19	2·16	2·09	2·05	2·01	1·98	1·93	1·87	1·83	1·74	1·69	1·65	65
2·33	2·28	2·23	2·19	2·16	2·13	2·07	2·02	1·98	1·95	1·90	1·84	1·80	1·71	1·66	1·62	70
2·29	2·23	2·19	2·15	2·11	2·08	2·02	1·97	1·93	1·90	1·85	1·79	1·75	1·66	1·60	1·56	80
2·25	2·20	2·15	2·12	2·08	2·05	1·99	1·94	1·90	1·87	1·82	1·75	1·71	1·62	1·56	1·52	90
2·23	2·17	2·13	2·09	2·05	2·02	1·96	1·91	1·87	1·84	1·79	1·72	1·68	1·59	1·53	1·49	100
2·18	2·13	2·08	2·04	2·01	1·98	1·91	1·86	1·82	1·79	1·74	1·67	1·63	1·53	1·47	1·42	125
2·15	2·10	2·05	2·01	1·98	1·94	1·88	1·83	1·79	1·76	1·70	1·63	1·59	1·49	1·42	1·37	150
2·11	2·06	2·01	1·97	1·94	1·91	1·84	1·79	1·75	1·71	1·66	1·59	1·54	1·44	1·37	1·31	200
2·07	2·02	1·97	1·93	1·90	1·87	1·80	1·75	1·71	1·67	1·61	1·54	1·50	1·39	1·31	1·25	300
2·04	1·99	1·94	1·90	1·87	1·84	1·77	1·72	1·67	1·64	1·58	1·51	1·46	1·35	1·26	1·18	500
2·02	1·97	1·92	1·88	1·84	1·81	1·75	1·69	1·65	1·61	1·56	1·48	1·43	1·31	1·22	1·13	1000
2·00	1·95	1·90	1·86	1·82	1·79	1·72	1·67	1·63	1·59	1·53	1·45	1·40	1·28	1·17	1·00	∞

DEGREES OF FREEDOM FOR THE DENOMINATOR (f_2)

Approximate formula for f_1 and f_2 larger than 30: $\log_{10} v_{.995}^2(f_1, f_2) \simeq \dfrac{2 \cdot 2373}{\sqrt{h - 1 \cdot 61}} - 1 \cdot 250 \left(\dfrac{1}{f_1} - \dfrac{1}{f_2} \right)$, where $\dfrac{1}{h} = \dfrac{1}{2} \left(\dfrac{1}{f_1} + \dfrac{1}{f_2} \right)$.

TABLE VII₈. 99·9 PER CENT FRACTILES OF THE v^2 DISTRIBUTION

$$v^2(f_1, f_2) = \frac{\chi_1^2/f_1}{\chi_2^2/f_2} = \frac{s_1^2}{s_2^2}.$$

DEGREES OF FREEDOM FOR THE NUMERATOR (f_1)

Multiply the numbers of the first row ($f_2 = 1$) by 1000.

f_2	1	2	3	4	5	6	7	8	9	10	15	20	30	50	100	200	500
1	405	500	540	562	576	586	593	598	602	606	616	621	626	630	633	635	636
2	998	999	999	999	999	999	999	999	999	999	999	999	999	999	999	999	999
3	168	148	141	137	135	133	132	131	130	129	127	126	125	125	124	124	124
4	74·1	61·2	56·2	53·4	51·7	50·5	49·7	49·0	48·5	48·0	46·8	46·1	45·4	44·9	44·5	44·3	44·1
5	47·0	36·6	33·2	31·1	29·8	28·8	28·2	27·6	27·2	26·9	25·9	25·4	24·9	24·4	24·1	23·9	23·8
6	35·5	27·0	23·7	21·9	20·8	20·0	19·5	19·0	18·7	18·4	17·6	17·1	16·7	16·3	16·0	15·9	15·8
7	29·2	21·7	18·8	17·2	16·2	15·5	15·0	14·6	14·3	14·1	13·3	12·9	12·5	12·2	11·9	11·8	11·7
8	25·4	18·5	15·8	14·4	13·5	12·9	12·4	12·0	11·8	11·5	10·8	10·5	10·1	9·80	9·57	9·46	9·39
9	22·9	16·4	13·9	12·6	11·7	11·1	10·7	10·4	10·1	9·89	9·24	8·90	8·55	8·26	8·04	7·93	7·86
10	21·0	14·9	12·6	11·3	10·5	9·92	9·52	9·20	8·96	8·75	8·13	7·80	7·47	7·19	6·98	6·87	6·81
11	19·7	13·8	11·6	10·4	9·58	9·05	8·66	8·35	8·12	7·92	7·32	7·01	6·68	6·41	6·21	6·10	6·04
12	18·6	13·0	10·8	9·63	8·89	8·38	8·00	7·71	7·48	7·29	6·71	6·40	6·09	5·83	5·63	5·52	5·46
13	17·8	12·3	10·2	9·07	8·35	7·86	7·49	7·21	6·98	6·80	6·23	5·93	5·62	5·37	5·17	5·07	5·01
14	17·1	11·8	9·73	8·62	7·92	7·43	7·08	6·80	6·58	6·40	5·85	5·56	5·25	5·00	4·80	4·70	4·64
15	16·6	11·3	9·34	8·25	7·57	7·09	6·74	6·47	6·26	6·08	5·53	5·25	4·95	4·70	4·51	4·41	4·35
16	16·1	11·0	9·00	7·94	7·27	6·81	6·46	6·19	5·98	5·81	5·27	4·99	4·70	4·45	4·26	4·16	4·10
17	15·7	10·7	8·73	7·68	7·02	6·56	6·22	5·96	5·75	5·58	5·05	4·78	4·48	4·24	4·05	3·95	3·89
18	15·4	10·4	8·49	7·46	6·81	6·35	6·02	5·76	5·56	5·39	4·87	4·59	4·30	4·06	3·87	3·77	3·71
19	15·1	10·2	8·28	7·26	6·61	6·18	5·84	5·59	5·39	5·22	4·70	4·43	4·14	3·90	3·71	3·61	3·55
20	14·8	9·95	8·10	7·10	6·46	6·02	5·69	5·44	5·24	5·08	4·56	4·29	4·01	3·77	3·58	3·48	3·42
22	14·4	9·61	7·80	6·81	6·19	5·76	5·44	5·19	4·99	4·83	4·32	4·06	3·77	3·53	3·34	3·25	3·19
24	14·0	9·34	7·55	6·59	5·98	5·55	5·23	4·99	4·80	4·64	4·14	3·87	3·59	3·35	3·16	3·07	3·01
26	13·7	9·12	7·36	6·41	5·80	5·38	5·07	4·83	4·64	4·48	3·99	3·72	3·45	3·20	3·01	2·92	2·86
28	13·5	8·93	7·19	6·25	5·66	5·24	4·93	4·69	4·50	4·35	3·86	3·60	3·32	3·08	2·89	2·79	2·73
30	13·3	8·77	7·05	6·12	5·53	5·12	4·82	4·58	4·39	4·24	3·75	3·49	3·22	2·98	2·79	2·69	2·63
40	12·6	8·25	6·60	5·70	5·13	4·73	4·43	4·21	4·02	3·87	3·40	3·15	2·87	2·64	2·44	2·34	2·28
50	12·2	7·95	6·34	5·46	4·90	4·51	4·22	4·00	3·82	3·67	3·20	2·95	2·68	2·44	2·24	2·14	2·07
60	12·0	7·76	6·17	5·31	4·76	4·37	4·09	3·87	3·69	3·54	3·08	2·83	2·56	2·31	2·11	2·01	1·93
80	11·7	7·54	5·97	5·13	4·58	4·21	3·92	3·70	3·53	3·39	2·93	2·68	2·40	2·16	1·95	1·84	1·77
100	11·5	7·41	5·85	5·01	4·48	4·11	3·83	3·61	3·44	3·30	2·84	2·59	2·32	2·07	1·87	1·75	1·68
200	11·2	7·15	5·64	4·81	4·29	3·92	3·65	3·43	3·26	3·12	2·67	2·42	2·15	1·90	1·68	1·55	1·46
500	11·0	7·01	5·51	4·69	4·18	3·82	3·54	3·33	3·16	3·02	2·58	2·33	2·05	1·80	1·57	1·43	1·32
∞	10·8	6·91	5·42	4·62	4·10	3·74	3·47	3·27	3·10	2·96	2·51	2·27	1·99	1·73	1·49	1·34	1·21

DEGREES OF FREEDOM FOR THE DENOMINATOR (f_2)

Example: $P\{v^2(8,20) < 5\cdot44\} = 99\cdot9\%.$

$v^2_{\cdot001}(f_1, f_2) = 1/v^2_{\cdot999}(f_2, f_1).$ Example: $v^2_{\cdot001}(8,20) = 1/v^2_{\cdot999}(20,8) = 1/10\cdot5 = 0\cdot095.$

Approximate formula for f_1 and f_2 larger than 30: $\log_{10} v^2_{\cdot999}(f_1, f_2) \simeq \dfrac{2\cdot6841}{\sqrt{h-2\cdot09}} - 1\cdot672\left(\dfrac{1}{f_1} - \dfrac{1}{f_2}\right)$, where $\dfrac{1}{h} = \dfrac{1}{2}\left(\dfrac{1}{f_1}\right.$

TABLE VII₉. 99·95 PER CENT FRACTILES OF THE v^2 DISTRIBUTION

$$v^2(f_1, f_2) = \frac{\chi_1^2/f_1}{\chi_2^2/f_2} = \frac{s_1^2}{s_2^2}.$$

DEGREES OF FREEDOM FOR THE NUMERATOR (f_1)

	1	2	3	4	5	6	7	8	9	10	15	20	30	50	100	200	500	∞
	Multiply the numbers of the first row ($f_2 = 1$) by 10,000 and the numbers of the second row ($f_2 = 2$) by 10.																	
1	162	200	216	225	231	234	237	239	241	242	246	248	250	252	253	253	254	254
2	200	200	200	200	200	200	200	200	200	200	200	200	200	200	200	200	200	200
3	266	237	225	218	214	211	209	208	207	206	203	201	199	198	197	197	196	196
4	106	87·4	80·1	76·1	73·6	71·9	70·6	69·7	68·9	68·3	66·5	65·5	64·6	63·8	63·2	62·9	62·7	62·6
5	63·6	49·8	44·4	41·5	39·7	38·5	37·6	36·9	36·4	35·9	34·6	33·9	33·1	32·5	32·1	31·8	31·7	31·6
6	46·1	34·8	30·4	28·1	26·6	25·6	24·9	24·3	23·9	23·5	22·4	21·9	21·4	20·9	20·5	20·3	20·2	20·1
7	37·0	27·2	23·5	21·4	20·2	19·3	18·7	18·2	17·8	17·5	16·5	16·0	15·5	15·1	14·7	14·6	14·5	14·4
8	31·6	22·8	19·4	17·6	16·4	15·7	15·1	14·6	14·3	14·0	13·1	12·7	12·2	11·8	11·6	11·4	11·4	11·3
9	28·0	19·9	16·8	15·1	14·1	13·3	12·8	12·4	12·1	11·8	11·0	10·6	10·2	9·80	9·53	9·40	9·32	9·26
10	25·5	17·9	15·0	13·4	12·4	11·8	11·3	10·9	10·6	10·3	9·56	9·16	8·75	8·42	8·16	8·04	7·96	7·90
11	23·6	16·4	13·6	12·2	11·2	10·6	10·1	9·76	9·48	9·24	8·52	8·14	7·75	7·43	7·18	7·06	6·98	6·93
12	22·2	15·3	12·7	11·2	10·4	9·74	9·28	8·94	8·66	8·43	7·74	7·37	7·00	6·68	6·45	6·33	6·25	6·20
13	21·1	14·4	11·9	10·5	9·66	9·07	8·63	8·29	8·03	7·81	7·13	6·78	6·42	6·11	5·88	5·76	5·69	5·64
14	20·2	13·7	11·3	9·95	9·11	8·53	8·11	7·78	7·52	7·31	6·65	6·31	5·95	5·66	5·43	5·31	5·24	5·19
15	19·5	13·2	10·8	9·48	8·66	8·10	7·68	7·36	7·11	6·91	6·27	5·93	5·58	5·29	5·06	4·94	4·87	4·83
16	18·9	12·7	10·3	9·08	8·29	7·74	7·33	7·02	6·77	6·57	5·94	5·61	5·27	4·98	4·76	4·64	4·57	4·52
17	18·4	12·3	9·99	8·75	7·98	7·44	7·04	6·73	6·49	6·29	5·67	5·34	5·01	4·72	4·50	4·39	4·32	4·27
18	17·9	11·9	9·69	8·47	7·71	7·18	6·78	6·48	6·24	6·05	5·44	5·12	4·78	4·50	4·28	4·17	4·10	4·06
19	17·5	11·6	9·42	8·23	7·48	6·95	6·57	6·27	6·03	5·84	5·25	4·92	4·59	4·31	4·09	3·98	3·91	3·87
20	17·2	11·4	9·20	8·02	7·28	6·76	6·38	6·08	5·85	5·66	5·07	4·75	4·42	4·15	3·93	3·82	3·75	3·70
22	16·6	11·0	8·82	7·67	6·94	6·44	6·07	5·78	5·55	5·36	4·79	4·47	4·15	3·88	3·66	3·55	3·48	3·44
24	16·2	10·6	8·52	7·39	6·68	6·18	5·82	5·54	5·31	5·13	4·55	4·25	3·93	3·66	3·44	3·33	3·27	3·22
26	15·8	10·3	8·27	7·16	6·46	5·98	5·62	5·34	5·12	4·94	4·37	4·07	3·75	3·48	3·27	3·16	3·09	3·04
28	15·5	10·1	8·07	6·98	6·28	5·80	5·45	5·18	4·96	4·78	4·22	3·92	3·61	3·34	3·13	3·01	2·95	2·90
30	15·2	9·90	7·90	6·82	6·14	5·66	5·31	5·04	4·82	4·65	4·10	3·80	3·48	3·22	3·00	2·89	2·82	2·78
40	14·4	9·25	7·33	6·30	5·64	5·19	4·85	4·59	4·38	4·21	3·68	3·39	3·08	2·82	2·60	2·49	2·41	2·37
50	13·9	8·88	7·01	6·01	5·37	4·93	4·60	4·34	4·14	3·98	3·45	3·16	2·86	2·59	2·37	2·25	2·17	2·13
60	13·6	8·65	6·81	5·82	5·20	4·76	4·44	4·18	3·98	3·82	3·30	3·02	2·71	2·45	2·23	2·11	2·03	1·98
80	13·2	8·37	6·57	5·60	4·99	4·56	4·24	4·00	3·80	3·65	3·13	2·85	2·54	2·27	2·05	1·92	1·84	1·79
100	13·0	8·21	6·43	5·47	4·87	4·44	4·13	3·89	3·70	3·54	3·03	2·75	2·44	2·18	1·95	1·82	1·74	1·67
200	12·5	7·90	6·16	5·23	4·64	4·23	3·92	3·68	3·49	3·34	2·83	2·56	2·25	1·98	1·74	1·60	1·50	1·42
500	12·3	7·72	6·01	5·09	4·51	4·10	3·80	3·56	3·36	3·21	2·72	2·45	2·14	1·87	1·61	1·46	1·34	1·24
∞	12·1	7·60	5·91	5·00	4·42	4·02	3·72	3·48	3·30	3·14	2·65	2·37	2·07	1·79	1·53	1·36	1·22	1·00

Example: $P\{v^2(8,20) < 6\cdot08\} = 99\cdot95\%$.

$v^2_{\cdot0005}(f_1, f_2) = 1/v^2_{\cdot9995}(f_2, f_1)$.   Example: $v^2_{\cdot0005}(8,20) = 1/v^2_{\cdot9995}(20,8) = 1/12\cdot7 = 0\cdot079$.

oximate formula for f_1 and f_2 larger than 30: $\log_{10} v^2_{\cdot9995}(f_1, f_2) \simeq \dfrac{2\cdot8580}{\sqrt{h-2\cdot30}} - 1\cdot857 \left(\dfrac{1}{f_1} - \dfrac{1}{f_2}\right)$, where $\dfrac{1}{h} = \dfrac{1}{2}\left(\dfrac{1}{f_1} + \dfrac{1}{f_2}\right)$.

TABLE VIII. THE DISTRIBUTION OF THE RANGE

Fractiles W_P of the Distribution of the Range.

Probability in per cent

n	$m\{W\}$ a_n	$\sigma\{W\}$ β_n	σ/m γ_n	0·05	0·1	0·5	1·0	2·5	5·0	10·0	20·0	30·0
2	1·128	·853	·756	0·00	0·00	0·01	0·02	0·04	0·09	0·18	0·36	0·55
3	1·693	·888	·525	0·04	0·06	0·13	0 19	0·30	0·43	0·62	0·90	1·14
4	2·059	·880	·427	0·16	0·20	0·34	0·43	0·59	0·76	0·98	1·29	1·53
5	2·326	·864	·371	0·31	0·37	0·55	0·66	0·85	1·03	1·26	1·57	1·82
6	2·534	·848	·335	0·47	0·54	0·75	0·87	1·06	1·25	1·49	1·80	2·04
7	2·704	·833	·308	0·61	0·69	0·92	1·05	1·25	1·44	1·68	1·99	2·22
8	2·847	·820	·288	0·75	0·83	1·08	1·20	1·41	1·60	1·83	2·14	2·38
9	2·970	·808	·272	0·88	0·96	1·21	1·34	1·55	1·74	1·97	2·28	2·51
10	3·078	·797	·259	1·00	1·08	1·33	1·47	1·67	1·86	2·09	2·39	2·62
11	3·173	·787	·248	1·10	1·20	1·45	1·58	1·78	1·97	2·20	2·50	2·72
12	3·258	·778	·239	1·21	1·30	1·55	1·68	1·88	2·07	2·30	2·59	2·82
13	3·336	·770	·231	1·30	1·39	1·64	1·77	1·97	2·16	2·39	2·68	2·90
14	3·407	·762	·224	1·38	1·48	1·72	1·86	2·06	2·24	2·47	2·75	2·97
15	3·472	·755	·217	1·46	1·56	1·80	1·93	2·14	2·32	2·54	2·83	3·04
16	3·532	·749	·212	1·53	1·63	1·88	2·01	2·21	2·39	2·61	2·89	3·11
17	3·588	·743	·207	1·60	1·69	1·94	2·07	2·27	2·45	2·67	2·95	3·17
18	3·640	·738	·203	1·66	1·75	2·01	2·14	2·34	2·51	2·73	3·01	3·22
19	3·689	·733	·199	1·72	1·82	2·07	2·20	2·39	2·57	2·79	3·06	3·27
20	3·735	·729	·195	1·78	1·88	2·12	2·25	2·45	2·63	2·84	3·11	3·32

Example: For $n = 10$ we get $m\{W\} = 3.078$, $\sigma\{W\} = 0.797$, and $P\{W < 4.79\} = 97.5\%$.

The range w denotes the difference between the largest and the smallest observation in a sample, $w = w_n = x_{(n)} - x_{(1)}$. The standardized range is $W = w/\sigma = u_{(n)} - u_{(1)}$. Table VIII contains $m\{W_n\} = a_n$, $\sigma\{W_n\} = \beta_n$, $\gamma_n = \beta_n/a_n$ and the fractiles W_P for $n = 2(1)20$. It follows that

$$m\{w_n\} = a_n\sigma. \quad \sigma\{w_n\} = \beta_n\sigma. \quad P\{w < \sigma W_P\} = P\left\{\frac{w}{W_P} < \sigma\right\} = P.$$

From k samples of n observations each from populations with the same variance we may compute the k ranges $w_1, \ldots, w_k$ and the mean range $\bar{w} = (w_1 + \ldots + w_k)/k$. It follows that $m\{\bar{w}\} = a_n\sigma$ and $\sigma\{\bar{w}\} = \beta_n\sigma/\sqrt{k}$. For values of n between 5 and 10, $\bar{w}$ is approximately normally distributed even for small values of k. Hence $\bar{w}/a_n$ will be an unbiased normally distributed estimate of σ with standard error $\gamma_n\sigma/\sqrt{k}$. Confidence limits for σ may be found as

$$P\left\{\frac{\dfrac{\bar{w}}{a_n}}{1 + u_{P_2}\dfrac{\gamma_n}{\sqrt{k}}} < \sigma < \frac{\dfrac{\bar{w}}{a_n}}{1 + u_{P_1}\dfrac{\gamma_n}{\sqrt{k}}}\right\} = P_2 - P_1.$$

TABLE VIII. THE DISTRIBUTION OF THE RANGE

FRACTILES W_P OF THE DISTRIBUTION OF THE RANGE.

PROBABILITY IN PER CENT

40.0	50.0	60.0	70.0	80.0	90.0	95.0	97.5	99.0	99.5	99.9	99.95	n
0·74	0·95	1·20	1·47	1·81	2·33	2·77	3·17	3·64	3·97	4·65	4·92	2
1·36	1·59	1·83	2·09	2·42	2·90	3·31	3·68	4·12	4·42	5·06	5·31	3
1·76	1·98	2·21	2·47	2·78	3·24	3·63	3·98	4·40	4·69	5·31	5·56	4
2·04	2·26	2·48	2·73	3·04	3·48	3·86	4·20	4·60	4·89	5·48	5·72	5
2·26	2·47	2·69	2·94	3·23	3·66	4·03	4·36	4·76	5·03	5·62	5·86	6
2·44	2·65	2·86	3·10	3·39	3·81	4·17	4·49	4·88	5·15	5·73	5·96	7
2·59	2·79	3·00	3·24	3·52	3·93	4·29	4·61	4·99	5·26	5·82	6·04	8
2·71	2·92	3·12	3·35	3·63	4·04	4·39	4·70	5·08	5·34	5·90	6·12	9
2·83	3·02	3·23	3·46	3·73	4·13	4·47	4·79	5·16	5·42	5·97	6·19	10
2·93	3·12	3·32	3·55	3·82	4·21	4·55	4·86	5·23	5·49	6·04	6·25	11
3·01	3·21	3·41	3·63	3·90	4·29	4·62	4·92	5·29	5·54	6·09	6·31	12
3·09	3·29	3·48	3·70	3·97	4·35	4·69	4·99	5·35	5·60	6·14	6·36	13
3·17	3·36	3·55	3·77	4·03	4·41	4·74	5·04	5·40	5·65	6·19	6·40	14
3·23	3·42	3·62	3·83	4·09	4·47	4·80	5·09	5·45	5·70	6·23	6·45	15
3·30	3·48	3·67	3·89	4·14	4·52	4·85	5·14	5·49	5·74	6·28	6·49	16
3·35	3·54	3·73	3·94	4·19	4·57	4·89	5·18	5·54	5·79	6·32	6·52	17
3·41	3·59	3·78	3·99	4·24	4·61	4·93	5·22	5·57	5·82	6·35	6·56	18
3·46	3·64	3·83	4·03	4·29	4·65	4·97	5·26	5·61	5·86	6·38	6·59	19
3·51	3·69	3·87	4·08	4·33	4·69	5·01	5·30	5·65	5·89	6·41	6·62	20

A table of a_n and a diagram for determining β_n may be found in *On the Extreme Individuals and the Range of Samples Taken From a Normal Population* by L. H. C. TIPPETT, Biometrika, 17, 1925, 364–387. A table of $P_n\{W\}$ for $n = 2(1)20$ and $W = 0·00(0·05)7·25$ has been given by E. S. PEARSON in *Biometrika*, 32, 1942, 301–308.

A large part of Table VIII is reproduced from the above-mentioned tables published in *Biometrika* by kind permission of the proprietors.

y	$z=f(y)$	y	$z=f(y)$	y	$z=f(y)$	y	$z=f(y)$	y	$z=f(y)$	y	$z=f(y)$	y	$z=f(y)$	y	$z=f(y)$
·550	−3·145	·600	−2·073	·650	−1·429	·700	−0·896	·750	−0·383	·800	0·168	·850	0·829	·900	1·749
·551	−3·113	·601	−2·058	·651	−1·418	·701	−0·886	·751	−0·372	·801	0·180	·851	0·844	·901	1·772
·552	−3·081	·602	−2·043	·652	−1·407	·702	−0·875	·752	−0·362	·802	0·192	·852	0·859	·902	1·795
·553	−3·050	·603	−2·028	·653	−1·395	·703	−0·865	·753	−0·351	·803	0·204	·853	0·874	·903	1·819
·554	−3·019	·604	−2·014	·654	−1·384	·704	−0·855	·754	−0·341	·804	0·216	·854	0·889	·904	1·843
·555	−2·990	·605	−1·999	·655	−1·373	·705	−0·845	·755	−0·330	·805	0·228	·855	0·905	·905	1·868
·556	−2·961	·606	−1·984	·656	−1·362	·706	−0·835	·756	−0·320	·806	0·240	·856	0·921	·906	1·892
·557	−2·933	·607	−1·970	·657	−1·351	·707	−0·824	·757	−0·309	·807	0·252	·857	0·936	·907	1·917
·558	−2·905	·608	−1·956	·658	−1·340	·708	−0·814	·758	−0·299	·808	0·264	·858	0·952	·908	1·943
·559	−2·878	·609	−1·942	·659	−1·329	·709	−0·804	·759	−0·288	·809	0·276	·859	0·968	·909	1·968
·560	−2·851	·610	−1·928	·660	−1·318	·710	−0·794	·760	−0·277	·810	0·289	·860	0·984	·910	1·994
·561	−2·826	·611	−1·914	·661	−1·307	·711	−0·783	·761	−0·267	·811	0·301	·861	1·000		
·562	−2·800	·612	−1·900	·662	−1·296	·712	−0·773	·762	−0·256	·812	0·313	·862	1·016		
·563	−2·775	·613	−1·886	·663	−1·285	·713	−0·763	·763	−0·245	·813	0·326	·863	1·033		
·564	−2·751	·614	−1·872	·664	−1·274	·714	−0·753	·764	−0·235	·814	0·338	·864	1·049		
·565	−2·727	·615	−1·859	·665	−1·263	·715	−0·743	·765	−0·224	·815	0·351	·865	1·066		
·566	−2·703	·616	−1·845	·666	−1·252	·716	−0·732	·766	−0·213	·816	0·363	·866	1·083		
·567	−2·680	·617	−1·832	·667	−1·242	·717	−0·722	·767	−0·203	·817	0·376	·867	1·100		
·568	−2·657	·618	−1·819	·668	−1·231	·718	−0·712	·768	−0·192	·818	0·388	·868	1·117		
·569	−2·635	·619	−1·806	·669	−1·220	·719	−0·702	·769	−0·181	·819	0·401	·869	1·134		
·570	−2·613	·620	−1·792	·670	−1·209	·720	−0·692	·770	−0·170	·820	0·414	·870	1·151		
·571	−2·591	·621	−1·779	·671	−1·199	·721	−0·681	·771	−0·159	·821	0·427	·871	1·168		
·572	−2·570	·622	−1·766	·672	−1·188	·722	−0·671	·772	−0·148	·822	0·439	·872	1·186		
·573	−2·549	·623	−1·754	·673	−1·177	·723	−0·661	·773	−0·137	·823	0·452	·873	1·204		
·574	−2·528	·624	−1·741	·674	−1·167	·724	−0·651	·774	−0·127	·824	0·466	·874	1·221		
·575	−2·508	·625	−1·728	·675	−1·156	·725	−0·641	·775	−0·116	·825	0·479	·875	1·240		
·576	−2·488	·626	−1·715	·676	−1·145	·726	−0·630	·776	−0·105	·826	0·492	·876	1·258		
·577	−2·468	·627	−1·703	·677	−1·135	·727	−0·620	·777	−0·094	·827	0·505	·877	1·276		
·578	−2·448	·628	−1·690	·678	−1·124	·728	−0·610	·778	−0·083	·828	0·518	·878	1·295		
·579	−2·429	·629	−1·678	·679	−1·114	·729	−0·600	·779	−0·071	·829	0·531	·879	1·313		
·580	−2·410	·630	−1·665	·680	−1·103	·730	−0·589	·780	−0·060	·830	0·545	·880	1·332		
·581	−2·391	·631	−1·653	·681	−1·093	·731	−0·579	·781	−0·049	·831	0·558	·881	1·351		
·582	−2·373	·632	−1·641	·682	−1·082	·732	−0·569	·782	−0·038	·832	0·572	·882	1·370		
·583	−2·355	·633	−1·629	·683	−1·072	·733	−0·559	·783	−0·027	·833	0·585	·883	1·389		
·584	−2·336	·634	−1·616	·684	−1·061	·734	−0·548	·784	−0·016	·834	0·599	·884	1·409		
·585	−2·319	·635	−1·604	·685	−1·051	·735	−0·538	·785	−0·004	·835	0·613	·885	1·428		
·586	−2·301	·636	−1·592	·686	−1·040	·736	−0·528	·786	+0·007	·836	0·627	·886	1·448		
·587	−2·284	·637	−1·580	·687	−1·030	·737	−0·517	·787	0·018	·837	0·641	·887	1·468		
·588	−2·266	·638	−1·568	·688	−1·020	·738	−0·507	·788	0·029	·838	0·655	·888	1·489		
·589	−2·249	·639	−1·557	·689	−1·009	·739	−0·497	·789	0·041	·839	0·669	·889	1·509		
·590	−2·232	·640	−1·545	·690	−0·999	·740	−0·487	·790	0·052	·840	0·683	·890	1·530		
·591	−2·216	·641	−1·533	·691	−0·989	·741	−0·476	·791	0·064	·841	0·697	·891	1·551		
·592	−2·199	·642	−1·521	·692	−0·978	·742	−0·466	·792	0·075	·842	0·711	·892	1·572		
·593	−2·183	·643	−1·510	·693	−0·968	·743	−0·455	·793	0·087	·843	0·726	·893	1·593		
·594	−2·167	·644	−1·498	·694	−0·958	·744	−0·445	·794	0·098	·844	0·740	·894	1·615		
·595	−2·151	·645	−1·486	·695	−0·947	·745	−0·435	·795	0·110	·845	0·754	·895	1·636		
·596	−2·135	·646	−1·475	·696	−0·937	·746	−0·424	·796	0·121	·846	0·769	·896	1·658		
·597	−2·119	·647	−1·463	·697	−0·927	·747	−0·414	·797	0·133	·847	0·784	·897	1·680		
·598	−2·104	·648	−1·452	·698	−0·917	·748	−0·404	·798	0·145	·848	0·799	·898	1·703		
·599	−2·089	·649	−1·441	·699	−0·906	·749	−0·393	·799	0·157	·849	0·814	·899	1·726		

TABLE IX. THE ONE-SIDED TRUNCATED NORMAL DISTRIBUTION   63

z	$g(z)$	δ	$\mu_{11}(z)$	$\mu_{12}(z)$	$\mu_{22}(z)$	$\varrho(z)$
−3·0	0·3328		1·015	−0·02415	0·5363	−0·033
		113				
−2·9	0·3441		1·019	−0·03120	0·5453	−0·042
		120				
−2·8	0·3561		1·026	−0·04001	0·5562	−0·053
		128				
−2·7	0·3689		1·034	−0·05094	0·5690	−0·066
		137				
−2·6	0·3826		1·044	−0·06445	0·5842	−0·083
		146				
−2·5	0·3972		1·057	−0·08105	0·6020	−0·102
		156				
−2·4	0·4128		1·074	−0·1013	0·6228	−0·124
		166				
−2·3	0·4294		1·096	−0·1261	0·6469	−0·150
		178				
−2·2	0·4472		1·124	−0·1562	0·6747	−0·179
		190				
−2·1	0·4662		1·159	−0·1927	0·7066	−0·213
		204				
−2·0	0·4866		1·204	−0·2367	0·7433	−0·250
		216				
−1·9	0·5082		1·260	−0·2899	0·7852	−0·291
		232				
−1·8	0·5314		1·332	−0·3538	0·8329	−0·336
		246				
−1·7	0·5560		1·424	−0·4305	0·8871	−0·383
		263				
−1·6	0·5823		1·540	−0·5226	0·9487	−0·432
		279				
−1·5	0·6102		1·688	−0·6327	1·018	−0·483
		296				
−1·4	0·6398		1·874	−0·7644	1·097	−0·533
		315				
−1·3	0·6713		2·110	−0·9215	1·186	−0·582
		332				
−1·2	0·7045		2·408	−1·109	1·287	−0·630
		351				
−1·1	0·7396		2·783	−1·331	1·400	−0·674
		370				
−1·0	0·7766		3·256	−1·596	1·527	−0·716
		390				
−0·9	0·8156		3·849	−1·910	1·671	−0·753
		409				
−0·8	0·8565		4·592	−2·281	1·831	−0·786
		428				
−0·7	0·8993		5·520	−2·719	2·012	−0·816
		449				
−0·6	0·9442		6·677	−3·236	2·214	−0·842
		467				
−0·5	0·9909		8·115	−3·845	2·440	−0·864
		487				
−0·4	1·0396		9·896	−4·559	2·692	−0·883
		506				
−0·3	1·0902		12·09	−5·397	2·975	−0·900
		526				
−0·2	1·1428		14·80	−6·376	3·290	−0·914
		543				
−0·1	1·1971		18·12	−7·520	3·641	−0·926
		562				
0·0	1·2533		22·19	−8·852	4·031	−0·936
		580				
0·1	1·3113		27·14	−10·40	4·465	−0·945
		597				
0·2	1·3710		33·16	−12·19	4·947	−0·952
		613				
0·3	1·4323		40·44	−14·27	5·481	−0·958
		630				
0·4	1·4953		49·23	−16·66	6·072	−0·964
		646				
0·5	1·5599		59·81	−19·42	6·725	−0·968
		660				
0·6	1·6259		72·49	−22·59	7·447	−0·972
		676				
0·7	1·6935		87·63	−26·22	8·242	−0·976
		689				
0·8	1·7624		105·7	−30·38	9·118	−0·979
		703				
0·9	1·8327		127·1	−35·12	10·08	−0·981
		716				
1·0	1·9043		152·4	−40·51	11·14	−0·983
		728				
1·1	1·9771		182·3	−46·65	12·30	−0·985
		740				
1·2	2·0511		217·4	−53·60	13·57	−0·987
		751				
1·3	2·1262		258·6	−61·47	14·96	−0·988
		762				
1·4	2·2024		306·8	−70·35	16·47	−0·990
		772				
1·5	2·2796		362·9	−80·35	18·12	−0·991
		782				
1·6	2·3578		428·1	−91·58	19·92	−0·992
		791				
1·7	2·4369		503·7	−104·2	21·88	−0·993
		800				
1·8	2·5169		591·1	−118·3	24·01	−0·993
		809				
1·9	2·5978		691·7	−134·1	26·31	−0·994
		816				
2·0	2·6794		807·6	−151·7	28·81	−0·995

y	$h=0.05$	0.10	0.15	0.20	0.25	0.30	0.35	0.40	0.45	0.50
0·500		−2·680	−1·985	−1·537	−1·207	−0·943	−0·720	−0·523	−0·345	−0·178
0·505		−2·555	−1·921	−1·499	−1·182	−0·925	−0·707	−0·514	−0·338	−0·173
0·510		−2·445	−1·862	−1·463	−1·158	−0·909	−0·695	−0·505	−0·331	−0·168
0·515		−2·349	−1·809	−1·429	−1·135	−0·893	−0·683	−0·496	−0·325	−0·163
0·520		−2·263	−1·759	−1·398	−1·114	−0·877	−0·672	−0·488	−0·318	−0·158
0·525	−2·922	−2·185	−1·714	−1·368	−1·093	−0·863	−0·661	−0·480	−0·312	−0·154
0·530	−2·786	−2·115	−1·671	−1·340	−1·074	−0·849	−0·651	−0·472	−0·306	−0·149
0·535	−2·667	−2·051	−1·632	−1·314	−1·055	−0·835	−0·641	−0·464	−0·300	−0·144
0·540	−2·562	−1·992	−1·595	−1·289	−1·037	−0·822	−0·631	−0·457	−0·295	−0·140
0·545	−2·468	−1·939	−1·560	−1·265	−1·020	−0·809	−0·621	−0·450	−0·289	−0·136
0·550	−2·384	−1·889	−1·528	−1·242	−1·004	−0·797	−0·612	−0·443	−0·284	−0·132
0·555	−2·308	−1·843	−1·497	−1·221	−0·988	−0·786	−0·603	−0·436	−0·278	−0·128
0·560	−2·239	−1·800	−1·468	−1·200	−0·973	−0·774	−0·595	−0·429	−0·273	−0·124
0·565	−2·175	−1·760	−1·441	−1·180	−0·959	−0·763	−0·587	−0·423	−0·268	−0·120
0·570	−2·117	−1·722	−1·415	−1·162	−0·945	−0·753	−0·579	−0·416	−0·263	−0·116
0·575	−2·063	−1·686	−1·390	−1·144	−0·932	−0·743	−0·571	−0·410	−0·258	−0·112
0·580	−2·013	−1·653	−1·366	−1·127	−0·919	−0·733	−0·563	−0·404	−0·254	−0·108
0·585	−1·966	−1·621	−1·344	−1·110	−0·906	−0·723	−0·555	−0·399	−0·249	−0·105
0·590	−1·923	−1·591	−1·322	−1·094	−0·894	−0·714	−0·548	−0·393	−0·245	−0·101
0·595	−1·882	−1·563	−1·302	−1·079	−0·882	−0·705	−0·541	−0·387	−0·240	−0·098
0·600	−1·844	−1·536	−1·282	−1·064	−0·871	−0·696	−0·534	−0·382	−0·236	−0·094
0·610	−1·773	−1·486	−1·245	−1·036	−0·850	−0·679	−0·521	−0·371	−0·228	−0·087
0·620	−1·710	−1·441	−1·211	−1·010	−0·829	−0·663	−0·508	−0·361	−0·220	−0·081
0·630	−1·654	−1·399	−1·180	−0·986	−0·810	−0·648	−0·496	−0·352	−0·212	−0·075
0·640	−1·602	−1·360	−1·150	−0·963	−0·792	−0·634	−0·485	−0·343	−0·204	−0·069
0·650	−1·555	−1·325	−1·123	−0·941	−0·775	−0·620	−0·474	−0·334	−0·197	−0·063
0·660	−1·512	−1·292	−1·097	−0·921	−0·759	−0·607	−0·463	−0·325	−0·190	−0·058
0·670	−1·472	−1·261	−1·073	−0·902	−0·744	−0·595	−0·453	−0·317	−0·184	−0·052
0·680	−1·435	−1·233	−1·050	−0·884	−0·729	−0·583	−0·444	−0·309	−0·177	−0·047
0·690	−1·401	−1·206	−1·029	−0·866	−0·715	−0·572	−0·434	−0·301	−0·171	−0·042
0·700	−1·369	−1·180	−1·009	−0·850	−0·702	−0·561	−0·425	−0·294	−0·166	−0·037
0·710	−1·339	−1·156	−0·989	−0·834	−0·689	−0·550	−0·417	−0·287	−0·160	−0·032
0·720	−1·311	−1·134	−0·971	−0·819	−0·677	−0·540	−0·408	−0·280	−0·154	−0·027
0·730	−1·285	−1·113	−0·954	−0·805	−0·665	−0·530	−0·400	−0·273	−0·148	−0·023
0·740	−1·260	−1·092	−0·937	−0·792	−0·654	−0·521	−0·393	−0·267	−0·143	−0·019
0·750	−1·236	−1·073	−0·921	−0·779	−0·643	−0·512	−0·385	−0·261	−0·138	−0·014
0·760	−1·214	−1·055	−0·906	−0·766	−0·632	−0·503	−0·378	−0·255	−0·133	−0·010
0·770	−1·193	−1·038	−0·892	−0·754	−0·622	−0·495	−0·371	−0·249	−0·128	−0·006
0·780	−1·173	−1·021	−0·878	−0·742	−0·613	−0·487	−0·364	−0·243	−0·123	−0·002
0·790	−1·154	−1·005	−0·865	−0·731	−0·603	−0·479	−0·357	−0·238	−0·118	0·002
0·800	−1·135	−0·990	−0·852	−0·721	−0·594	−0·471	−0·351	−0·232	−0·114	0·006
0·850	−1·055	−0·922	−0·795	−0·673	−0·553	−0·437	−0·322	−0·207	−0·093	0·023
0·900	−0·989	−0·866	−0·747	−0·631	−0·518	−0·407	−0·296	−0·186	−0·074	0·039
0·950	−0·934	−0·819	−0·706	−0·596	−0·488	−0·380	−0·273	−0·166	−0·058	0·053
1·000	−0·887	−0·778	−0·671	−0·565	−0·461	−0·357	−0·253	−0·149	−0·043	0·066
1·050	−0·847	−0·742	−0·639	−0·538	−0·437	−0·336	−0·235	−0·133	−0·029	0·077
1·100	−0·811	−0·711	−0·611	−0·513	−0·415	−0·317	−0·219	−0·119	−0·017	0·088
1·150	−0·779	−0·682	−0·586	−0·491	−0·396	−0·300	−0·203	−0·105	−0·005	0·098
1·200	−0·751	−0·657	−0·564	−0·471	−0·378	−0·284	−0·189	−0·093	0·006	0·107
1·250	−0·725	−0·634	−0·544	−0·453	−0·362	−0·270	−0·177	−0·082	0·016	0·116
1·300	−0·702	−0·613	−0·525	−0·436	−0·347	−0·257	−0·165	−0·072	0·025	0·124
1·350	−0·680	−0·594	−0·507	−0·421	−0·333	−0·244	−0·154	−0·062	0·033	0·131
1·400	−0·661	−0·576	−0·491	−0·406	−0·320	−0·233	−0·144	−0·053	0·041	0·138
1·450	−0·643	−0·560	−0·477	−0·393	−0·308	−0·222	−0·134	−0·044	0·049	0·145
1·500	−0·626	−0·545	−0·463	−0·381	−0·297	−0·212	−0·125	−0·036	0·056	0·151

TABLE X. THE ONE-SIDED CENSORED NORMAL DISTRIBUTION   65

z	$\psi'(z)$	δ	δ^2	$\mu_{11}(z)$	$\mu_{12}(z)$	$\mu_{22}(z)$	$\varrho(z)$
−3.0	3.2831	−928		1.000	$-0.0^{3}3348$	0.5012	−0.000
−2.9	3.1903	−924	4	1.000	$-0.0^{3}4721$	0.5016	−0.001
−2.8	3.0979	−921	3	1.000	$-0.0^{3}6599$	0.5022	−0.001
−2.7	3.0058	−917	4	1.000	$-0.0^{3}9144$	0.5030	−0.001
−2.6	2.9141	−914	3	1.000	$-0.0^{2}1256$	0.5040	−0.002
−2.5	2.8227	−909	5	1.001	$-0.0^{2}1712$	0.5052	−0.002
−2.4	2.7318	−904	5	1.001	$-0.0^{2}2312$	0.5069	−0.003
−2.3	2.6414	−899	5	1.001	$-0.0^{2}3099$	0.5090	−0.004
−2.2	2.5515	−894	5	1.001	$-0.0^{2}4121$	0.5117	−0.006
−2.1	2.4621	−889	5	1.002	$-0.0^{2}5438$	0.5149	−0.008
−2.0	2.3732	−883	6	1.003	$-0.0^{2}7123$	0.5190	−0.010
−1.9	2.2849	−876	7	1.004	$-0.0^{2}9265$	0.5239	−0.013
−1.8	2.1973	−870	6	1.005	−0.01197	0.5299	−0.016
−1.7	2.1103	−862	8	1.006	−0.01537	0.5371	−0.021
−1.6	2.0241	−854	8	1.009	−0.01961	0.5458	−0.026
−1.5	1.9387	−846	8	1.011	−0.02488	0.5562	−0.033
−1.4	1.8541	−837	9	1.015	−0.03141	0.5685	−0.041
−1.3	1.7704	−828	9	1.019	−0.03946	0.5830	−0.051
−1.2	1.6876	−818	10	1.025	−0.04936	0.6000	−0.063
−1.1	1.6058	−807	11	1.032	−0.06149	0.6200	−0.077
−1.0	1.5251	−795	12	1.042	−0.07634	0.6434	−0.093
−0.9	1.4456	−782	13	1.054	−0.09450	0.6707	−0.112
−0.8	1.3674	−769	13	1.069	−0.1167	0.7025	−0.135
−0.7	1.2905	−755	14	1.089	−0.1437	0.7395	−0.160
−0.6	1.2150	−739	16	1.114	−0.1768	0.7826	−0.189
−0.5	1.1411	−723	16	1.147	−0.2172	0.8327	−0.222
−0.4	1.0688	−706	17	1.189	−0.2666	0.8911	−0.259
−0.3	0.99817	−6875	19	1.243	−0.3271	0.9592	−0.300
−0.2	0.92942	−6680	195	1.312	−0.4013	1.039	−0.344
−0.1	0.86262	−6474	206	1.401	−0.4926	1.132	−0.391
0.0	0.79788	−6255	219	1.517	−0.6052	1.241	−0.441
0.1	0.73533	−6026	229	1.667	−0.7445	1.370	−0.492
0.2	0.67507	−5785	241	1.863	−0.9172	1.523	−0.545
0.3	0.61722	−5534	251	2.118	−1.132	1.704	−0.596
0.4	0.56188	−5272	262	2.453	−1.401	1.919	−0.646
0.5	0.50916	−5001	271	2.893	−1.738	2.178	−0.692
0.6	0.45915	−4723	278	3.473	−2.162	2.488	−0.735
0.7	0.41192	−4436	287	4.241	−2.699	2.863	−0.744
0.8	0.36756	−4145	291	5.261	−3.381	3.319	−0.809
0.9	0.32611	−3851	294	6.623	−4.252	3.876	−0.839
1.0	0.28760	−3555	296	8.448	−5.370	4.561	−0.865
1.1	0.25205	−3261	294	10.90	−6.812	5.408	−0.887
1.2	0.21944	−2970	291	14.22	−8.682	6.462	−0.906
1.3	0.18974	−2686	284	18.73	−11.12	7.780	−0.921
1.4	0.16288	−2409	277	24.89	−14.32	9.442	−0.934
1.5	0.13879	−2144	265	33.34	−18.54	11.55	−0.945
1.6	0.11735	−1891	253	44.99	−24.14	14.24	−0.954
1.7	0.098436	−16543	237	61.13	−31.62	17.71	−0.961
1.8	0.081893	−14337	2206	83.64	−41.66	22.19	−0.967
1.9	0.067556	−12308	2029	115.2	−55.25	28.05	−0.972
2.0	0.055248			159.7	−73.75	35.74	−0.976

$$g(h, z) = \cfrac{1}{\cfrac{h}{1-h}\psi'(z)-z} = \frac{n-a}{a\psi'(z)-(n-a)z}, \quad \text{where} \quad h = \frac{a}{n}.$$

TABLE XI. TWO-SIDED 95 PER CENT CONFIDENCE LIM[...]

x DENOTES THE NUMERATOR AND n[...]

DENOMINATOR MINUS NUMERATOR

NUMERATOR OF THE RELATIVE FREQUENCY

x \ $n-x$	1	2	3	4	5	6	7	8	9	10	11	12	13	14	15	16
0	975/000	842/000	708/000	602/000	522/000	459/000	410/000	369/000	336/000	308/000	285/000	265/000	247/000	232/000	218/000	206/000
1	987/013	906/008	806/006	716/005	641/004	579/004	527/003	483/003	445/003	413/002	385/002	360/002	339/002	319/002	302/002	287/001
2	992/094	932/068	853/053	777/043	710/037	651/032	600/028	556/025	518/023	484/021	454/019	428/018	405/017	383/016	364/015	347/014
3	994/194	947/147	882/118	816/099	755/085	701/075	652/067	610/060	572/055	538/050	508/047	481/043	456/040	434/038	414/036	396/034
4	995/284	957/223	901/184	843/157	788/137	738/122	692/109	651/099	614/091	581/084	551/078	524/073	499/068	476/064	456/061	437/057
5	996/359	963/290	915/245	863/212	813/187	766/167	723/151	684/139	649/128	616/118	587/110	560/103	535/097	512/091	491/087	471/082
6	996/421	968/349	925/299	878/262	833/234	789/211	749/192	711/177	677/163	646/152	617/142	590/133	565/126	543/119	522/113	502/107
7	997/473	972/400	933/348	891/308	849/277	808/251	770/230	734/213	701/198	671/184	643/173	616/163	592/154	570/146	549/139	529/132
8	997/517	975/444	940/390	901/349	861/316	823/289	787/266	753/247	722/230	692/215	665/203	639/191	616/181	593/172	573/164	553/156
9	997/555	977/482	945/428	909/386	872/351	837/323	802/299	770/278	740/260	711/244	685/231	660/218	636/207	615/197	594/188	575/180
10	998/587	979/516	950/462	916/419	882/384	848/354	816/329	785/308	756/289	728/272	702/257	678/244	655/232	634/221	614/211	595/202
11	998/615	981/546	953/492	922/449	890/413	858/383	827/357	797/335	769/315	743/298	718/282	694/268	672/256	651/244	631/234	612/224
12	998/640	982/572	957/519	927/476	897/440	867/410	837/384	809/361	782/340	756/322	732/306	709/291	687/278	666/266	647/255	628/245
13	998/661	983/595	960/544	932/501	903/465	874/435	846/408	819/384	793/364	768/345	744/328	722/313	701/299	680/287	661/275	643/264
14	998/681	984/617	962/566	936/524	909/488	881/457	854/430	828/407	803/385	779/366	756/349	734/334	713/320	694/306	675/295	657/283
15	998/698	985/636	964/586	939/544	913/509	887/478	861/451	836/427	812/406	789/386	766/369	745/353	725/339	705/325	687/313	669/302
16	999/713	986/653	966/604	943/563	918/529	893/498	868/471	844/447	820/425	798/405	776/388	755/372	736/357	717/343	698/331	681/319
17	999/727	987/669	968/621	946/581	922/547	898/516	874/488	851/465	828/443	806/423	785/406	765/389	745/374	727/360	709/347	692/335
18	999/740	988/683	970/637	948/597	925/564	902/533	879/506	857/482	835/460	814/440	793/422	773/406	755/391	736/376	719/363	702/351
19	999/751	988/696	971/651	950/612	929/579	906/549	884/522	862/498	841/476	821/456	801/439	782/422	763/406	745/392	728/379	712/366
20	999/762	989/708	972/664	953/626	932/593	910/564	889/537	868/513	847/492	827/472	808/454	789/437	771/421	753/407	737/393	720/381
22	999/781	990/730	975/688	956/651	937/619	917/590	897/565	877/541	858/519	839/500	820/481	803/465	785/449	768/434	752/421	737/408
24	999/797	991/749	976/708	960/673	942/642	923/614	904/589	885/566	867/545	849/525	831/507	814/490	798/475	782/460	766/446	751/433
26	999/810	991/765	978/726	962/693	945/663	928/636	910/611	893/588	875/567	858/548	841/530	825/513	809/497	794/483	779/469	764/456
28	999/822	992/779	980/743	965/710	949/681	932/655	916/631	899/609	882/588	866/569	850/551	834/535	819/519	804/504	790/491	776/478
30	999/833	992/792	981/757	967/725	952/697	936/672	920/649	904/627	889/607	873/588	858/571	843/554	828/539	814/524	800/510	786/498
35	999/855	993/818	983/786	971/758	958/732	944/708	930/686	916/666	902/647	888/628	874/612	860/596	847/581	834/567	821/554	809/541
40	999/871	994/838	985/809	975/783	963/759	951/737	938/717	925/698	912/679	900/662	887/646	875/631	862/616	850/602	838/590	827/578
45	999/885	995/855	987/828	977/804	967/782	956/761	944/742	933/724	921/707	909/690	898/675	886/661	875/647	864/633	853/621	842/609
50	1000/896	995/868	988/843	979/821	970/800	960/781	949/763	939/746	928/730	917/714	906/700	896/686	885/673	875/660	865/648	854/636
60	1000/912	996/888	990/867	983/848	975/830	966/813	957/797	948/782	939/767	929/752	920/740	911/727	902/715	893/703	884/692	874/681
80	1000/933	997/915	992/898	987/882	981/868	974/855	967/842	960/829	953/816	945/804	938/793	931/783	923/773	916/763	909/753	901/744
100	1000/946	998/931	994/917	989/904	984/892	979/881	973/870	967/859	962/849	955/838	949/829	943/820	937/811	931/802	925/794	919/786
200	1000/973	999/965	997/957	995/951	992/944	989/938	986/932	983/926	980/920	977/914	974/909	970/903	967/898	964/893	961/888	957/883
500	1000/989	1000/986	999/983	998/980	997/977	996/974	995/972	993/969	992/967	991/964	989/962	988/960	986/957	985/955	984/953	982/950
∞	1000/1000	1000/1000	1000/1000	1000/1000	1000/1000	1000/1000	1000/1000	1000/1000	1000/1000	1000/1000	1000/1000	1000/1000	1000/1000	1000/1000	1000/1000	1000/1000

Example: Observed relative frequency 8/30 = [...]

OMINATOR OF THE RELATIVE FREQUENCY

DENOMINATOR MINUS NUMERATOR

19	20	22	24	26	28	30	35	40	45	50	60	80	100	200	500	∞	$n-x$ / x
176 000	168 000	154 000	142 000	132 000	123 000	116 000	100 000	088 000	079 000	071 000	060 000	045 000	036 000	018 000	007 000	000 000	0
249 001	238 001	219 001	203 001	190 001	178 001	167 001	145 001	129 001	115 001	104 001	088 000	067 000	054 000	027 000	011 000	000 000	1
304 012	292 011	270 010	251 009	235 009	221 008	208 008	182 007	162 006	145 005	132 005	112 004	085 003	069 002	035 001	014 000	000 000	2
349 029	336 028	312 025	292 024	274 022	257 020	243 019	214 017	191 015	172 013	157 012	133 010	102 008	083 006	043 003	017 001	000 000	3
388 050	374 047	349 044	327 040	307 038	290 035	275 033	242 029	217 025	196 023	179 021	152 017	118 013	096 011	049 005	020 002	000 000	4
421 071	407 068	381 063	358 058	337 055	319 051	303 048	268 042	241 037	218 033	200 030	170 025	132 019	108 016	056 008	023 003	000 000	5
451 094	436 090	410 083	386 077	364 072	345 068	328 064	292 056	263 049	239 044	219 040	187 034	145 026	119 021	062 011	026 004	000 000	6
478 116	463 111	435 103	411 096	389 090	369 084	351 080	314 070	283 062	258 056	237 051	203 043	158 033	130 027	068 014	028 005	000 000	7
502 138	487 132	459 123	434 115	412 107	391 101	373 096	334 084	302 075	276 067	254 061	218 052	171 040	141 033	074 017	031 007	000 000	8
524 159	508 153	481 142	455 133	433 125	412 118	393 111	353 098	321 088	293 079	270 072	233 061	184 047	151 038	080 020	033 008	000 000	9
544 179	528 173	500 161	475 151	452 142	431 134	412 127	372 112	338 100	310 091	286 083	248 071	196 055	162 045	086 023	036 009	000 000	10
561 199	546 192	519 180	493 169	470 159	449 150	429 142	388 126	354 113	325 102	300 094	260 080	207 062	171 051	091 026	038 011	000 000	11
578 218	563 211	535 197	510 186	487 175	465 166	446 157	404 140	369 125	339 114	314 104	273 089	217 069	180 057	097 030	040 012	000 000	12
594 237	579 229	551 215	525 202	503 191	481 181	461 172	419 153	384 138	353 125	327 115	285 098	227 077	189 063	102 033	043 014	000 000	13
608 255	593 247	566 232	540 218	517 206	496 196	476 186	433 166	398 150	367 136	340 125	297 107	237 084	198 069	107 036	045 015	000 000	14
621 272	607 263	579 248	554 234	531 221	509 210	490 200	446 179	410 162	379 147	352 135	308 116	247 091	206 075	112 039	047 016	000 000	15
634 288	619 280	592 263	567 249	544 236	522 224	502 214	459 191	422 173	391 158	364 146	319 126	256 099	214 081	117 043	050 018	000 000	16
645 304	631 295	604 278	579 263	556 250	535 238	515 227	471 203	434 185	402 169	375 156	330 134	266 106	222 087	122 046	052 019	000 000	17
656 319	642 310	615 293	590 277	568 264	547 251	527 240	483 215	445 196	413 179	386 165	340 143	274 113	230 093	127 050	054 021	000 000	18
666 334	652 324	626 307	601 291	578 277	557 264	538 252	494 227	456 207	424 189	396 175	350 152	283 120	238 099	132 053	056 022	000 000	19
676 348	662 338	636 320	612 304	589 289	568 276	548 264	504 238	467 217	434 199	406 184	359 160	292 126	245 105	137 057	059 024	000 000	20
693 374	680 364	654 346	631 329	614 314	588 300	568 287	524 260	487 237	454 219	425 203	378 177	308 140	260 117	146 063	063 027	000 000	22
709 399	696 388	671 369	648 352	626 337	605 322	586 309	543 281	505 257	472 237	443 220	395 193	324 154	274 128	155 070	067 030	000 000	24
723 422	711 411	686 386	663 374	642 358	622 343	603 330	559 300	522 276	489 255	460 237	411 208	338 167	287 140	164 077	072 033	000 000	26
736 443	724 432	700 412	678 395	657 378	637 363	618 349	575 319	538 294	505 272	475 254	426 223	353 180	300 153	172 083	076 036	000 000	28
748 462	736 452	713 432	691 414	670 397	651 382	632 368	590 337	552 311	520 289	490 269	441 237	366 192	313 162	181 090	080 039	000 000	30
773 506	762 496	740 476	719 457	700 441	681 425	663 410	622 378	586 351	553 327	524 306	474 272	397 222	342 188	201 106	090 046	000 000	35
793 544	783 533	763 513	743 495	724 478	706 462	689 448	649 414	614 386	581 361	553 340	503 303	425 250	368 213	220 122	099 053	000 000	40
811 576	801 566	781 546	763 528	745 511	728 495	711 480	673 447	639 419	607 393	579 370	529 332	451 276	392 236	238 137	109 061	000 000	45
825 604	816 594	797 575	780 557	763 540	746 525	731 510	694 476	660 447	630 421	602 398	552 359	474 301	415 259	255 152	118 068	000 000	50
848 650	840 641	823 622	807 605	792 589	777 574	763 559	728 526	697 497	668 471	641 448	593 407	515 345	455 300	287 181	136 083	000 000	60
880 717	874 708	860 692	846 676	833 662	820 647	808 634	778 603	750 575	724 549	699 526	655 485	580 420	520 370	342 234	169 111	000 000	80
901 762	895 755	883 740	872 726	860 713	847 700	838 687	812 658	787 632	764 608	741 585	700 545	630 480	571 429	395 280	199 138	000 000	100
947 868	943 863	937 854	930 845	923 836	917 828	910 819	894 799	878 780	863 762	848 745	819 713	766 658	720 605	550 450	319 253	000 000	200
978 944	976 941	973 937	970 933	967 928	964 924	961 920	954 910	947 901	939 891	932 882	917 864	889 831	862 801	747 681	531 469	000 000	500
1000 1000	1000 1000	1000 1000	1000 1000	1000 1000	1000 1000	1000 1000	1000 1000	1000 1000	1000 1000	1000 1000	1000 1000	1000 1000	1000 1000	1000 1000	1000 1000	— —	∞

NUMERATOR OF THE RELATIVE FREQUENCY

5 per cent confidence limits for θ are 0·123 and 0·459.

TABLE XI. TWO-SIDED 99 PER CENT CONFIDENCE LIMITS

x DENOTES THE NUMERATOR AND n T...

NUMERATOR OF THE RELATIVE FREQUENCY (rows) — **DENOMINATOR MINUS NUMERATOR** (columns)

Each cell shows upper limit / lower limit.

x \ $n-x$	1	2	3	4	5	6	7	8	9	10	11	12	13	14	15	16
0	995/000	929/000	829/000	734/000	653/000	586/000	531/000	484/000	445/000	411/000	382/000	357/000	335/000	315/000	298/000	282/000
1	997/003	959/002	889/001	815/001	746/001	685/001	632/001	585/001	544/001	509/000	477/000	449/000	424/000	402/000	381/000	363/000
2	998/041	971/029	917/023	856/019	797/016	742/014	693/012	648/011	608/010	573/009	541/008	512/008	486/007	463/007	441/006	422/006
3	999/111	977/083	934/066	882/055	830/047	781/042	735/037	693/033	655/030	621/028	589/026	561/024	534/022	510/021	488/020	468/019
4	999/185	981/144	945/118	900/100	854/087	809/077	767/069	728/062	691/057	658/053	627/049	599/045	573/043	549/040	527/038	507/036
5	999/254	984/203	953/170	913/146	872/128	831/114	791/103	755/094	720/087	688/080	658/075	631/070	605/065	582/062	560/058	539/055
6	999/315	986/258	958/219	923/191	886/169	848/152	811/138	777/127	744/117	714/109	685/101	658/095	633/090	610/085	588/080	567/076
7	999/368	988/307	963/265	931/233	897/209	862/189	828/172	795/159	764/147	735/137	707/128	681/121	657/114	634/108	612/102	592/097
8	999/415	989/352	967/307	938/272	906/245	873/223	841/205	811/189	781/176	753/165	726/155	701/146	677/138	655/131	634/125	614/119
9	999/456	990/392	970/345	943/309	913/280	883/256	853/236	824/219	795/205	768/192	743/181	718/171	695/162	674/154	653/146	634/140
10	1000/491	991/427	972/379	947/342	920/312	891/286	863/265	835/247	808/232	782/218	758/205	734/195	712/185	690/176	670/168	651/161
11	1000/523	992/459	974/411	951/373	925/342	899/315	872/293	845/274	819/257	795/242	771/229	748/218	726/207	705/197	686/189	667/181
12	1000/551	992/488	976/439	955/401	930/369	905/342	879/319	854/299	829/282	805/266	782/252	760/240	739/228	719/218	700/209	682/200
13	1000/576	993/514	978/466	957/427	935/395	910/367	886/343	862/323	838/305	815/288	793/274	772/261	751/249	731/238	713/228	695/219
14	1000/598	993/537	979/490	960/451	938/418	915/390	892/366	869/345	846/326	824/310	803/295	782/281	762/269	743/257	724/247	707/237
15	1000/619	994/559	980/512	962/473	942/440	920/412	898/388	875/366	854/347	832/330	811/314	791/300	772/287	753/276	735/265	718/255
16	1000/637	994/578	981/532	964/493	945/461	924/433	903/408	881/386	860/366	839/349	819/333	800/318	781/305	763/293	745/282	728/272
17	1000/654	994/596	982/551	966/512	947/480	927/452	907/427	887/405	866/385	846/367	827/351	808/336	789/322	772/309	754/298	738/288
18	1000/669	995/613	983/568	968/530	950/498	931/469	911/445	891/422	872/402	852/384	833/368	815/353	797/339	780/326	763/315	747/304
19	1000/683	995/628	984/584	969/547	952/515	934/486	915/462	896/439	877/419	858/401	840/384	822/369	804/355	787/342	771/330	755/319
20	1000/696	995/642	985/599	971/562	954/530	936/502	918/478	900/455	881/435	863/417	845/400	828/384	811/370	794/357	778/345	763/334
22	1000/719	996/668	986/626	973/590	958/559	941/531	924/507	907/484	890/464	873/445	856/429	839/413	823/398	807/385	792/372	777/361
24	1000/738	996/690	987/649	975/615	961/584	946/557	930/533	913/511	897/490	881/471	865/455	849/439	834/424	819/410	804/398	789/386
26	1000/755	996/709	988/670	977/637	963/607	949/580	934/557	919/535	903/515	888/496	873/479	858/463	843/448	829/434	814/4-2	800/410
28	1000/770	996/726	989/689	978/656	966/627	952/602	938/578	924/557	909/537	894/518	880/501	866/485	852/471	838/457	824/444	811/432
30	1000/784	997/741	989/705	980/674	968/646	955/621	942/598	928/577	914/557	900/539	886/522	873/506	859/492	846/478	833/464	820/452
35	1000/811	997/773	991/740	982/711	972/685	961/661	949/639	937/619	924/600	912/582	900/566	887/550	875/535	863/522	851/510	839/498
40	1000/832	998/797	992/767	984/740	975/716	965/694	955/673	944/654	933/636	921/619	910/603	899/588	888/574	876/560	865/548	854/536
45	1000/849	998/817	993/790	986/765	978/742	969/721	959/701	949/683	939/666	929/649	919/634	908/620	898/606	888/593	877/582	867/570
50	1000/863	998/834	994/808	987/785	980/763	972/743	963/725	954/708	945/691	935/676	926/661	916/643	906/634	897/621	888/610	878/599
60	1000/884	998/859	995/836	989/816	983/797	976/780	969/763	961/748	953/733	945/719	937/705	928/693	920/680	912/668	904/657	895/646
80	1000/912	999/892	996/874	992/857	987/842	982/827	976/814	970/801	964/788	957/776	951/765	944/754	938/743	931/733	924/724	918/715
100	1000/929	999/912	997/897	993/884	990/871	985/858	981/847	976/836	971/825	965/815	960/805	955/795	949/786	943/777	938/769	932/761
200	1000/964	999/955	998/947	997/939	995/932	992/925	990/919	988/913	985/907	982/901	979/896	976/890	973/884	970/878	967/873	964/868
500	1000/985	1000/982	999/978	999/975	998/972	997/969	996/967	995/964	994/961	993/959	992/956	990/953	989/951	988/949	987/946	985/944
∞	1000/1000	1000/1000	1000/1000	1000/1000	1000/1000	1000/1000	1000/1000	1000/1000	1000/1000	1000/1000	1000/1000	1000/1000	1000/1000	1000/1000	1000/1000	1000/1000

Example: Observed relative frequency 8/30 = 0·...

NOMINATOR OF THE RELATIVE FREQUENCY

DENOMINATOR MINUS NUMERATOR

19	20	22	24	26	28	30	35	40	45	50	60	80	100	200	500	∞	x
243 *000*	233 *000*	214 *000*	198 *000*	184 *000*	172 *000*	162 *000*	140 *000*	124 *000*	111 *000*	101 *000*	085 *000*	064 *000*	052 *000*	026 *000*	011 *000*	000 *000*	0
317 *000*	304 *000*	281 *000*	262 *000*	245 *000*	230 *000*	216 *000*	189 *000*	168 *000*	151 *000*	137 *000*	116 *000*	088 *000*	071 *000*	036 *000*	015 *000*	000 *000*	1
372 *005*	358 *005*	332 *004*	310 *004*	291 *004*	274 *004*	259 *003*	227 *003*	203 *002*	183 *002*	166 *002*	141 *002*	108 *001*	088 *001*	045 *001*	018 *000*	000 *000*	2
416 *016*	401 *015*	374 *014*	351 *013*	330 *012*	311 *011*	295 *011*	260 *009*	233 *008*	210 *007*	192 *006*	164 *005*	126 *004*	103 *003*	053 *002*	022 *001*	000 *000*	3
453 *031*	438 *029*	410 *027*	385 *025*	363 *023*	344 *022*	326 *020*	289 *018*	260 *016*	235 *014*	215 *013*	184 *011*	143 *008*	116 *007*	061 *003*	025 *001*	000 *000*	4
485 *048*	470 *046*	441 *042*	416 *039*	393 *037*	373 *034*	354 *032*	315 *028*	284 *025*	258 *022*	237 *020*	203 *017*	158 *013*	129 *010*	068 *005*	028 *002*	000 *000*	5
514 *066*	498 *064*	469 *059*	443 *054*	420 *051*	398 *048*	379 *045*	339 *039*	306 *035*	279 *031*	257 *028*	220 *024*	173 *018*	142 *015*	075 *008*	031 *003*	000 *000*	6
538 *085*	522 *082*	493 *076*	467 *070*	443 *066*	422 *062*	402 *058*	361 *051*	327 *045*	299 *041*	275 *037*	237 *031*	186 *024*	153 *019*	081 *010*	033 *004*	000 *000*	7
561 *104*	545 *100*	516 *093*	489 *087*	465 *081*	443 *076*	423 *072*	381 *063*	346 *056*	317 *051*	292 *046*	252 *039*	199 *030*	164 *024*	087 *012*	036 *005*	000 *000*	8
581 *123*	565 *119*	536 *110*	510 *103*	485 *097*	463 *091*	443 *086*	400 *076*	364 *067*	334 *061*	309 *055*	267 *047*	212 *036*	175 *029*	093 *015*	039 *006*	000 *000*	9
599 *142*	583 *137*	555 *127*	529 *119*	504 *112*	482 *106*	461 *100*	418 *088*	381 *079*	351 *071*	324 *065*	281 *055*	224 *043*	185 *035*	099 *018*	041 *007*	000 *000*	10
616 *160*	600 *155*	571 *144*	545 *135*	521 *127*	499 *120*	478 *114*	434 *100*	397 *090*	366 *081*	339 *074*	295 *063*	235 *049*	195 *040*	104 *021*	044 *008*	000 *000*	11
631 *178*	616 *172*	587 *161*	561 *151*	537 *142*	515 *134*	494 *127*	450 *113*	412 *101*	380 *092*	352 *084*	307 *072*	246 *056*	205 *045*	110 *024*	047 *010*	000 *000*	12
645 *196*	630 *189*	602 *177*	576 *166*	552 *157*	529 *148*	508 *141*	465 *125*	426 *112*	394 *102*	366 *094*	320 *080*	257 *062*	214 *051*	116 *027*	049 *011*	000 *000*	13
658 *213*	643 *206*	615 *193*	590 *181*	566 *171*	543 *162*	522 *154*	478 *137*	440 *124*	407 *112*	379 *103*	332 *088*	267 *069*	223 *057*	122 *030*	051 *012*	000 *000*	14
670 *229*	655 *222*	628 *208*	602 *196*	578 *186*	556 *176*	536 *167*	490 *149*	452 *135*	418 *123*	390 *112*	343 *096*	276 *076*	231 *062*	127 *033*	054 *013*	000 *000*	15
681 *245*	666 *237*	639 *223*	614 *211*	590 *200*	568 *189*	548 *180*	502 *161*	464 *146*	430 *133*	401 *122*	354 *105*	285 *082*	239 *068*	132 *036*	056 *015*	000 *000*	16
692 *260*	677 *252*	650 *238*	625 *225*	602 *213*	580 *202*	559 *193*	514 *173*	475 *156*	442 *143*	412 *131*	364 *113*	295 *089*	247 *073*	137 *039*	059 *016*	000 *000*	17
701 *275*	687 *267*	661 *252*	636 *238*	612 *226*	591 *215*	570 *205*	525 *184*	486 *167*	453 *153*	423 *141*	374 *122*	304 *096*	255 *079*	142 *042*	061 *018*	000 *000*	18
711 *289*	697 *281*	670 *265*	646 *251*	623 *239*	601 *228*	581 *217*	536 *195*	497 *177*	463 *162*	433 *150*	384 *130*	313 *102*	263 *084*	147 *045*	063 *019*	000 *000*	19
719 *303*	705 *295*	679 *279*	655 *264*	632 *251*	611 *239*	591 *229*	546 *206*	507 *187*	473 *172*	443 *158*	394 *137*	321 *109*	271 *090*	152 *048*	066 *020*	000 *000*	20
735 *330*	721 *321*	696 *304*	673 *289*	650 *274*	629 *263*	609 *251*	565 *227*	526 *207*	492 *190*	462 *176*	411 *153*	337 *122*	286 *101*	162 *054*	070 *023*	000 *000*	22
749 *354*	736 *345*	711 *327*	688 *312*	666 *298*	646 *285*	626 *273*	582 *247*	543 *226*	509 *208*	479 *193*	428 *168*	353 *134*	300 *112*	171 *061*	075 *026*	000 *000*	24
761 *377*	749 *368*	726 *350*	702 *334*	681 *319*	661 *306*	642 *293*	598 *267*	560 *244*	526 *225*	495 *209*	444 *183*	368 *147*	313 *122*	180 *067*	079 *029*	000 *000*	26
772 *399*	761 *389*	737 *371*	715 *354*	694 *339*	675 *325*	656 *312*	613 *285*	575 *262*	541 *242*	510 *225*	459 *198*	382 *159*	326 *133*	189 *073*	083 *031*	000 *000*	28
783 *419*	771 *409*	749 *391*	727 *374*	707 *358*	688 *344*	669 *331*	626 *303*	589 *278*	555 *258*	525 *241*	473 *212*	395 *171*	339 *143*	197 *079*	088 *034*	000 *000*	30
805 *464*	794 *454*	773 *435*	753 *418*	733 *402*	715 *387*	697 *374*	657 *343*	620 *318*	587 *296*	557 *277*	505 *245*	426 *200*	368 *169*	218 *095*	098 *041*	000 *000*	35
823 *503*	813 *493*	793 *474*	774 *457*	756 *440*	738 *425*	722 *411*	682 *380*	646 *354*	614 *330*	585 *310*	534 *276*	453 *227*	394 *193*	237 *110*	108 *048*	000 *000*	40
838 *537*	828 *527*	810 *508*	792 *491*	775 *474*	758 *459*	742 *445*	704 *413*	670 *386*	638 *362*	609 *341*	559 *305*	478 *253*	418 *216*	255 *125*	118 *055*	000 *000*	45
850 *567*	842 *557*	824 *538*	807 *521*	791 *505*	775 *490*	759 *475*	723 *443*	690 *415*	659 *391*	631 *369*	581 *332*	501 *277*	440 *238*	273 *139*	127 *062*	000 *000*	50
870 *616*	863 *606*	847 *589*	832 *572*	817 *556*	802 *541*	788 *527*	755 *495*	724 *466*	695 *441*	668 *419*	620 *380*	541 *321*	479 *278*	305 *167*	145 *076*	000 *000*	60
898 *687*	891 *679*	878 *663*	866 *647*	853 *632*	841 *618*	829 *605*	800 *574*	773 *547*	747 *522*	723 *499*	679 *459*	604 *396*	543 *349*	360 *219*	179 *103*	000 *000*	80
916 *737*	910 *729*	899 *714*	888 *700*	878 *687*	867 *674*	857 *661*	831 *632*	807 *606*	784 *582*	762 *560*	722 *521*	651 *457*	593 *407*	407 *265*	209 *129*	000 *000*	100
955 *853*	952 *848*	946 *838*	939 *829*	933 *820*	927 *811*	921 *803*	905 *782*	890 *763*	875 *745*	861 *727*	833 *695*	781 *640*	735 *593*	565 *435*	332 *243*	000 *000*	200
981 *937*	980 *934*	977 *930*	974 *925*	971 *921*	969 *917*	966 *912*	959 *902*	952 *892*	945 *882*	938 *873*	924 *855*	897 *821*	871 *791*	757 *668*	541 *459*	000 *000*	500
1000 *1000*	1000 *1000*	1000 *1000*	1000 *1000*	1000 *1000*	1000 *1000*	1000 *1000*	1000 *1000*	1000 *1000*	1000 *1000*	1000 *1000*	1000 *1000*	1000 *1000*	1000 *1000*	1000 *1000*	1000 *1000*	— 	∞

NUMERATOR OF THE RELATIVE FREQUENCY

99 per cent confidence limits for θ are 0·093 and 0·516.

TABLE XII. $y = 2 \arcsin \sqrt{x}$.

	·000	·001	·002	·003	·004	·005	·006	·007	·008	·009
·00	0·0000	0·0633	0·0895	0·1096	0·1266	0·1415	0·1551	0·1675	0·1791	0·1900
·01	0·2003	0·2101	0·2195	0·2285	0·2372	0·2456	0·2537	0·2615	0·2691	0·2766
·02	0·2838	0·2909	0·2977	0·3045	0·3111	0·3176	0·3239	0·3301	0·3362	0·3423
·03	0·3482	0·3540	0·3597	0·3653	0·3709	0·3764	0·3818	0·3871	0·3924	0·3976
·04	0·4027	0·4078	0·4128	0·4178	0·4227	0·4275	0·4323	0·4371	0·4418	0·4464
·05	0·4510	0·4556	0·4601	0·4646	0·4690	0·4734	0·4778	0·4822	0·4864	0·4907
·06	0·4949	0·4991	0·5033	0·5074	0·5115	0·5156	0·5196	0·5236	0·5276	0·5316
·07	0·5355	0·5394	0·5433	0·5472	0·5510	0·5548	0·5586	0·5624	0·5661	0·5698
·08	0·5735	0·5772	0·5808	0·5845	0·5881	0·5917	0·5953	0·5988	0·6024	0·6059
·09	0·6094	0·6129	0·6163	0·6198	0·6232	0·6266	0·6300	0·6334	0·6368	0·6402
·10	0·6435	0·6468	0·6501	0·6534	0·6567	0·6600	0·6632	0·6665	0·6697	0·6729
·11	0·6761	0·6793	0·6825	0·6857	0·6888	0·6920	0·6951	0·6982	0·7013	0·7044
·12	0·7075	0·7106	0·7136	0·7167	0·7197	0·7227	0·7258	0·7288	0·7318	0·7347
·13	0·7377	0·7407	0·7437	0·7466	0·7495	0·7525	0·7554	0·7583	0·7612	0·7641
·14	0·7670	0·7699	0·7727	0·7756	0·7785	0·7813	0·7841	0·7870	0·7898	0·7926
·15	0·7954	0·7982	0·8010	0·8038	0·8065	0·8093	0·8121	0·8148	0·8176	0·8203
·16	0·8230	0·8258	0·8285	0·8312	0·8339	0·8366	0·8393	0·8420	0·8446	0·8473
·17	0·8500	0·8526	0·8553	0·8579	0·8606	0·8632	0·8658	0·8685	0·8711	0·8737
·18	0·8763	0·8789	0·8815	0·8841	0·8867	0·8892	0·8918	0·8944	0·8969	0·8995
·19	0·9021	0·9046	0·9071	0·9097	0·9122	0·9147	0·9173	0·9198	0·9223	0·9248
·20	0·9273	0·9298	0·9323	0·9348	0·9373	0·9397	0·9422	0·9447	0·9471	0·9496
·21	0·9521	0·9545	0·9570	0·9594	0·9619	0·9643	0·9667	0·9692	0·9716	0·9740
·22	0·9764	0·9788	0·9812	0·9836	0·9860	0·9884	0·9908	0·9932	0·9956	0·9980
·23	1·0004	1·0027	1·0051	1·0075	1·0098	1·0122	1·0146	1·0169	1·0193	1·0216
·24	1·0239	1·0263	1·0286	1·0310	1·0333	1·0356	1·0379	1·0403	1·0426	1·0449
·25	1·0472	1·0495	1·0518	1·0541	1·0564	1·0587	1·0610	1·0633	1·0656	1·0679
·26	1·0701	1·0724	1·0747	1·0770	1·0792	1·0815	1·0838	1·0860	1·0883	1·0905
·27	1·0928	1·0951	1·0973	1·0995	1·1018	1·1040	1·1063	1·1085	1·1107	1·1130
·28	1·1152	1·1174	1·1196	1·1219	1·1241	1·1263	1·1285	1·1307	1·1329	1·1351
·29	1·1373	1·1396	1·1418	1·1440	1·1461	1·1483	1·1505	1·1527	1·1549	1·1571
·30	1·1593	1·1615	1·1636	1·1658	1·1680	1·1702	1·1723	1·1745	1·1767	1·1788
·31	1·1810	1·1832	1·1853	1·1875	1·1896	1·1918	1·1939	1·1961	1·1982	1·2004
·32	1·2025	1·2047	1·2068	1·2090	1·2111	1·2132	1·2154	1·2175	1·2196	1·2217
·33	1·2239	1·2260	1·2281	1·2303	1·2324	1·2345	1·2366	1·2387	1·2408	1·2430
·34	1·2451	1·2472	1·2493	1·2514	1·2535	1·2556	1·2577	1·2598	1·2619	1·2640
·35	1·2661	1·2682	1·2703	1·2724	1·2745	1·2766	1·2787	1·2807	1·2828	1·2849
·36	1·2870	1·2891	1·2912	1·2932	1·2953	1·2974	1·2995	1·3016	1·3036	1·3057
·37	1·3078	1·3098	1·3119	1·3140	1·3161	1·3181	1·3202	1·3222	1·3243	1·3264
·38	1·3284	1·3305	1·3325	1·3346	1·3367	1·3387	1·3408	1·3428	1·3449	1·3469
·39	1·3490	1·3510	1·3531	1·3551	1·3572	1·3592	1·3613	1·3633	1·3654	1·3674
·40	1·3694	1·3715	1·3735	1·3756	1·3776	1·3796	1·3817	1·3837	1·3857	1·3878
·41	1·3898	1·3918	1·3939	1·3959	1·3979	1·4000	1·4020	1·4040	1·4061	1·4081
·42	1·4101	1·4121	1·4142	1·4162	1·4182	1·4202	1·4222	1·4243	1·4263	1·4283
·43	1·4303	1·4324	1·4344	1·4364	1·4384	1·4404	1·4424	1·4445	1·4465	1·4485
·44	1·4505	1·4525	1·4545	1·4565	1·4586	1·4606	1·4626	1·4646	1·4666	1·4686
·45	1·4706	1·4726	1·4746	1·4767	1·4787	1·4807	1·4827	1·4847	1·4867	1·4887
·46	1·4907	1·4927	1·4947	1·4967	1·4987	1·5007	1·5027	1·5048	1·5068	1·5088
·47	1·5108	1·5128	1·5148	1·5168	1·5188	1·5208	1·5228	1·5248	1·5268	1·5288
·48	1·5308	1·5328	1·5348	1·5368	1·5388	1·5408	1·5428	1·5448	1·5468	1·5488
·49	1·5508	1·5528	1·5548	1·5568	1·5588	1·5608	1·5628	1·5648	1·5668	1·5688

Example: $2 \arcsin \sqrt{0.296} = 1.1505$.

TABLE XII. $y = 2 \arcsin \sqrt{x}$.

	·000	·001	·002	·003	·004	·005	·006	·007	·008	·009
·50	1·5708	1·5728	1·5748	1·5768	1·5788	1·5808	1·5828	1·5848	1·5868	1·5888
·51	1·5908	1·5928	1·5948	1·5968	1·5988	1·6008	1·6028	1·6048	1·6068	1·6088
·52	1·6108	1·6128	1·6148	1·6168	1·6188	1·6208	1·6228	1·6248	1·6268	1·6288
·53	1·6308	1·6328	1·6348	1·6368	1·6388	1·6409	1·6429	1·6449	1·6469	1·6489
·54	1·6509	1·6529	1·6549	1·6569	1·6589	1·6609	1·6629	1·6649	1·6669	1·6690
·55	1·6710	1·6730	1·6750	1·6770	1·6790	1·6810	1·6830	1·6850	1·6871	1·6891
·56	1·6911	1·6931	1·6951	1·6971	1·6992	1·7012	1·7032	1·7052	1·7072	1·7092
·57	1·7113	1·7133	1·7153	1·7173	1·7193	1·7214	1·7234	1·7254	1·7274	1·7295
·58	1·7315	1·7335	1·7355	1·7376	1·7396	1·7416	1·7437	1·7457	1·7477	1·7498
·59	1·7518	1·7538	1·7559	1·7579	1·7599	1·7620	1·7640	1·7660	1·7681	1·7701
·60	1·7722	1·7742	1·7762	1·7783	1·7803	1·7824	1·7844	1·7865	1·7885	1·7906
·61	1·7926	1·7947	1·7967	1·7988	1·8008	1·8029	1·8049	1·8070	1·8090	1·8111
·62	1·8132	1·8152	1·8173	1·8193	1·8214	1·8235	1·8255	1·8276	1·8297	1·8317
·63	1·8338	1·8359	1·8380	1·8400	1·8421	1·8442	1·8463	1·8483	1·8504	1·8525
·64	1·8546	1·8567	1·8588	1·8608	1·8629	1·8650	1·8671	1·8692	1·8713	1·8734
·65	1·8755	1·8776	1·8797	1·8818	1·8839	1·8860	1·8881	1·8902	1·8923	1·8944
·66	1·8965	1·8986	1·9008	1·9029	1·9050	1·9071	1·9092	1·9113	1·9135	1·9156
·67	1·9177	1·9198	1·9220	1·9241	1·9262	1·9284	1·9305	1·9326	1·9348	1·9369
·68	1·9391	1·9412	1·9434	1·9455	1·9477	1·9498	1·9520	1·9541	1·9563	1·9584
·69	1·9606	1·9628	1·9649	1·9671	1·9693	1·9714	1·9736	1·9758	1·9780	1·9801
·70	1·9823	1·9845	1·9867	1·9889	1·9911	1·9932	1·9954	1·9976	1·9998	2·0020
·71	2·0042	2·0064	2·0087	2·0109	2·0131	2·0153	2·0175	2·0197	2·0219	2·0242
·72	2·0264	2·0286	2·0309	2·0331	2·0353	2·0376	2·0398	2·0420	2·0443	2·0465
·73	2·0488	2·0510	2·0533	2·0556	2·0578	2·0601	2·0624	2·0646	2·0669	2·0692
·74	2·0714	2·0737	2·0760	2·0783	2·0806	2·0829	2·0852	2·0875	2·0898	2·0921
·75	2·0944	2·0967	2·0990	2·1013	2·1037	2·1060	2·1083	2·1106	2·1130	2·1153
·76	2·1176	2·1200	2·1223	2·1247	2·1270	2·1294	2·1318	2·1341	2·1365	2·1389
·77	2·1412	2·1436	2·1460	2·1484	2·1508	2·1532	2·1556	2·1580	2·1604	2·1628
·78	2·1652	2·1676	2·1700	2·1724	2·1749	2·1773	2·1797	2·1822	2·1846	2·1871
·79	2·1895	2·1920	2·1944	2·1969	2·1994	2·2019	2·2043	2·2068	2·2093	2·2118
·80	2·2143	2·2168	2·2193	2·2218	2·2243	2·2269	2·2294	2·2319	2·2345	2·2370
·81	2·2395	2·2421	2·2446	2·2472	2·2498	2·2523	2·2549	2·2575	2·2601	2·2627
·82	2·2653	2·2679	2·2705	2·2731	2·2758	2·2784	2·2810	2·2837	2·2863	2·2890
·83	2·2916	2·2943	2·2970	2·2996	2·3023	2·3050	2·3077	2·3104	2·3131	2·3158
·84	2·3186	2·3213	2·3240	2·3268	2·3295	2·3323	2·3351	2·3378	2·3406	2·3434
·85	2·3462	2·3490	2·3518	2·3546	2·3575	2·3603	2·3631	2·3660	2·3689	2·3717
·86	2·3746	2·3775	2·3804	2·3833	2·3862	2·3891	2·3921	2·3950	2·3979	2·4009
·87	2·4039	2·4068	2·4098	2·4128	2·4158	2·4189	2·4219	2·4249	2·4280	2·4310
·88	2·4341	2·4372	2·4403	2·4434	2·4465	2·4496	2·4528	2·4559	2·4591	2·4623
·89	2·4655	2·4687	2·4719	2·4751	2·4783	2·4816	2·4849	2·4882	2·4915	2·4948
·90	2·4981	2·5014	2·5048	2·5082	2·5115	2·5149	2·5184	2·5218	2·5253	2·5287
·91	2·5322	2·5357	2·5392	2·5428	2·5463	2·5499	2·5535	2·5571	2·5607	2·5644
·92	2·5681	2·5718	2·5755	2·5792	2·5830	2·5868	2·5906	2·5944	2·5983	2·6022
·93	2·6061	2·6100	2·6140	2·6179	2·6220	2·6260	2·6301	2·6342	2·6383	2·6425
·94	2·6467	2·6509	2·6551	2·6594	2·6638	2·6681	2·6725	2·6770	2·6815	2·6860
·95	2·6906	2·6952	2·6998	2·7045	2·7093	2·7141	2·7189	2·7238	2·7288	2·7338
·96	2·7389	2·7440	2·7492	2·7545	2·7598	2·7652	2·7707	2·7762	2·7819	2·7876
·97	2·7934	2·7993	2·8053	2·8115	2·8177	2·8240	2·8305	2·8371	2·8438	2·8507
·98	2·8578	2·8650	2·8725	2·8801	2·8879	2·8960	2·9044	2·9131	2·9221	2·9314
·99	2·9413	2·9516	2·9625	2·9741	2·9865	3·0001	3·0150	3·0320	3·0521	3·0783
1·00	3·1416									

Example: $2 \arcsin \sqrt{0.724} = 2.0353$.

TABLE XIII. LOGARITHMS OF $n!$

n	$\log n!$	n	$\log n!$	n	$\log n!$	n	$\log n!$	n	$\log n!$
1	0·0000	51	66·1906	101	159·9743	151	264·9359	201	377·2001
2	0·3010	52	67·9066	102	161·9829	152	267·1177	202	379·5054
3	0·7782	53	69·6309	103	163·9958	153	269·3024	203	381·8129
4	1·3802	54	71·3633	104	166·0128	154	271·4899	204	384·1226
5	2·0792	55	73·1037	105	168·0340	155	273·6803	205	386·4343
6	2·8573	56	74·8519	106	170·0593	156	275·8734	206	388·7482
7	3·7024	57	76·6077	107	172·0887	157	278·0693	207	391·0642
8	4·6055	58	78·3712	108	174·1221	158	280·2679	208	393·3822
9	5·5598	59	80·1420	109	176·1595	159	282·4693	209	395·7024
10	6·5598	60	81·9202	110	178·2009	160	284·6735	210	398·0246
11	7·6012	61	83·7055	111	180·2462	161	286·8803	211	400·3489
12	8·6803	62	85·4979	112	182·2955	162	289·0898	212	402·6752
13	9·7943	63	87·2972	113	184·3485	163	291·3020	213	405·0036
14	10·9404	64	89·1034	114	186·4054	164	293·5168	214	407·3340
15	12·1165	65	90·9163	115	188·4661	165	295·7343	215	409·6664
16	13·3206	66	92·7359	116	190·5306	166	297·9544	216	412·0009
17	14·5511	67	94·5619	117	192·5988	167	300·1771	217	414·3373
18	15·8063	68	96·3945	118	194·6707	168	302·4024	218	416·6758
19	17·0851	69	98·2333	119	196·7462	169	304·6303	219	419·0162
20	18·3861	70	100·0784	120	198·8254	170	306·8608	220	421·3587
21	19·7083	71	101·9297	121	200·9082	171	309·0938	221	423·7031
22	21·0508	72	103·7870	122	202·9945	172	311·3293	222	426·0494
23	22·4125	73	105·6503	123	205·0844	173	313·5674	223	428·3977
24	23·7927	74	107·5196	124	207·1779	174	315·8079	224	430·7480
25	25·1906	75	109·3946	125	209·2748	175	318·0509	225	433·1002
26	26·6056	76	111·2754	126	211·3751	176	320·2965	226	435·4543
27	28·0370	77	113·1619	127	213·4790	177	322·5444	227	437·8103
28	29·4841	78	115·0540	128	215·5862	178	324·7948	228	440·1682
29	30·9465	79	116·9516	129	217·6967	179	327·0477	229	442·5281
30	32·4237	80	118·8547	130	219·8107	180	329·3030	230	444·8898
31	33·9150	81	120·7632	131	221·9280	181	331·5606	231	447·2534
32	35·4202	82	122·6770	132	224·0485	182	333·8207	232	449·6189
33	36·9387	83	124·5961	133	226·1724	183	336·0832	233	451·9862
34	38·4702	84	126·5204	134	228·2995	184	338·3480	234	454·3555
35	40·0142	85	128·4498	135	230·4298	185	340·6152	235	456·7265
36	41·5705	86	130·3843	136	232·5634	186	342·8847	236	459·0994
37	43·1387	87	132·3238	137	234·7001	187	345·1565	237	461·4742
38	44·7185	88	134·2683	138	236·8400	188	347·4307	238	463·8508
39	46·3096	89	136·2177	139	238·9830	189	349·7071	239	466·2292
40	47·9116	90	138·1719	140	241·1291	190	351·9859	240	468·6094
41	49·5244	91	140·1310	141	243·2783	191	354·2669	241	470·9914
42	51·1477	92	142·0948	142	245·4306	192	356·5502	242	473·3752
43	52·7811	93	144·0632	143	247·5860	193	358·8358	243	475·7608
44	54·4246	94	146·0364	144	249·7443	194	361·1236	244	478·1482
45	56·0778	95	148·0141	145	251·9057	195	363·4136	245	480·5374
46	57·7406	96	149·9964	146	254·0700	196	365·7059	246	482·9283
47	59·4127	97	151·9831	147	256·2374	197	368·0003	247	485·3210
48	61·0939	98	153·9744	148	258·4076	198	370·2970	248	487·7154
49	62·7841	99	155·9700	149	260·5808	199	372·5959	249	490·1116
50	64·4831	100	157·9700	150	262·7569	200	374·8969	250	492·5096

TABLE XIII. LOGARITHMS OF $n!$

n	log $n!$	n	log $n!$	n	log $n!$	n	log $n!$	n	log $n!$
251	494·9093	301	616·9644	351	742·6373	401	871·4096	451	1002·8931
252	497·3107	302	619·4444	352	745·1838	402	874·0138	452	1005·5482
253	499·7138	303	621·9258	353	747·7316	403	876·6191	453	1008·2043
254	502·1186	304	624·4087	354	750·2806	404	879·2255	454	1010·8614
255	504·5252	305	626·8930	355	752·8308	405	881·8329	455	1013·5194
256	506·9334	306	629·3787	356	755·3823	406	884·4415	456	1016·1783
257	509·3433	307	631·8659	357	757·9349	407	887·0510	457	1018·8383
258	511·7549	308	634·3544	358	760·4888	408	889·6617	458	1021·4991
259	514·1682	309	636·8444	359	763·0439	409	892·2734	459	1024·1609
260	516·5832	310	639·3357	360	765·6002	410	894·8862	460	1026·8237
261	518·9999	311	641·8285	361	768·1577	411	897·5001	461	1029·4874
262	521·4182	312	644·3226	362	770·7164	412	900·1150	462	1032·1520
263	523·8381	313	646·8182	363	773·2764	413	902·7309	463	1034·8176
264	526·2597	314	649·3151	364	775·8375	414	905·3479	464	1037·4841
265	528·6830	315	651·8134	365	778·3997	415	907·9660	465	1040·1516
266	531·1079	316	654·3131	366	780·9632	416	910·5850	466	1042·8200
267	533·5344	317	656·8142	367	783·5279	417	913·2052	467	1045·4893
268	535·9625	318	659·3166	368	786·0937	418	915·8264	468	1048·1595
269	538·3922	319	661·8204	369	788·6608	419	918·4486	469	1050·8307
270	540·8236	320	664·3255	370	791·2290	420	921·0718	470	1053·5028
271	543·2566	321	666·8320	371	793·7983	421	923·6961	471	1056·1758
272	545·6912	322	669·3399	372	796·3689	422	926·3214	472	1058·8498
273	548·1273	323	671·8491	373	798·9406	423	928·9478	473	1061·5246
274	550·5651	324	674·3596	374	801·5135	424	931·5751	474	1064·2004
275	553·0044	325	676·8715	375	804·0875	425	934·2035	475	1066·8771
276	555·4453	326	679·3847	376	806·6627	426	936·8329	476	1069·5547
277	557·8878	327	681·8993	377	809·2390	427	939·4633	477	1072·2332
278	560·3318	328	684·4152	378	811·8165	428	942·0948	478	1074·9127
279	562·7774	329	686·9324	379	814·3952	429	944·7272	479	1077·5930
280	565·2246	330	689·4509	380	816·9749	430	947·3607	480	1080·2742
281	567·6733	331	691·9707	381	819·5559	431	949·9952	481	1082·9564
282	570·1235	332	694·4918	382	822·1379	432	952·6307	482	1085·6394
283	572·5753	333	697·0143	383	824·7211	433	955·2672	483	1088·3234
284	575·0287	334	699·5380	384	827·3055	434	957·9047	484	1091·0082
285	577·4835	335	702·0631	385	829·8909	435	960·5431	485	1093·6940
286	579·9399	336	704·5894	386	832·4775	436	963·1826	486	1096·3806
287	582·3977	337	707·1170	387	835·0652	437	965·8231	487	1099·0681
288	584·8571	338	709·6460	388	837·6540	438	968·4646	488	1101·7565
289	587·3180	339	712·1762	389	840·2440	439	971·1071	489	1104·4458
290	589·7804	340	714·7076	390	842·8351	440	973·7505	490	1107·1360
291	592·2443	341	717·2404	391	845·4272	441	976·3949	491	1109·8271
292	594·7097	342	719·7744	392	848·0205	442	979·0404	492	1112·5191
293	597·1766	343	722·3097	393	850·6149	443	981·6868	493	1115·2119
294	599·6449	344	724·8463	394	853·2104	444	984·3342	494	1117·9057
295	602·1147	345	727·3841	395	855·8070	445	986·9825	495	1120·6003
296	604·5860	346	729·9232	396	858·4047	446	989·6318	496	1123·2958
297	607·0588	347	732·4635	397	861·0035	447	992·2822	497	1125·9921
298	609·5330	348	735·0051	398	863·6034	448	994·9334	498	1128·6893
299	612·0087	349	737·5479	399	866·2044	449	997·5857	499	1131·3874
300	614·4858	350	740·0920	400	868·8064	450	1000·2389	500	1134·0864

TABLE XIII. LOGARITHMS OF $n!$

n	$\log n!$	n	$\log n!$	n	$\log n!$	n	$\log n!$	n	$\log n!$
501	1136·7862	551	1272·8480	601	1410·8812	651	1550·7215	701	1692·2299
502	1139·4870	552	1275·5899	602	1413·6608	652	1553·5357	702	1695·0762
503	1142·1885	553	1278·3327	603	1416·4411	653	1556·3506	703	1697·9232
504	1144·8909	554	1281·0762	604	1419·2221	654	1559·1662	704	1700·7708
505	1147·5942	555	1283·8205	605	1422·0039	655	1561·9824	705	1703·6190
506	1150·2984	556	1286·5655	606	1424·7863	656	1564·7993	706	1706·4678
507	1153·0034	557	1289·3114	607	1427·5695	657	1567·6169	707	1709·3172
508	1155·7093	558	1292·0580	608	1430·3534	658	1570·4351	708	1712·1672
509	1158·4160	559	1294·8054	609	1433·1380	659	1573·2540	709	1715·0179
510	1161·1236	560	1297·5536	610	1435·9234	660	1576·0736	710	1717·8691
511	1163·8320	561	1300·3026	611	1438·7094	661	1578·8938	711	1720·7210
512	1166·5412	562	1303·0523	612	1441·4962	662	1581·7146	712	1723·5735
513	1169·2514	563	1305·8028	613	1444·2836	663	1584·5361	713	1726·4266
514	1171·9623	564	1308·5541	614	1447·0718	664	1587·3583	714	1729·2803
515	1174·6741	565	1311·3062	615	1449·8607	665	1590·1811	715	1732·1346
516	1177·3868	566	1314·0590	616	1452·6503	666	1593·0046	716	1734·9895
517	1180·1003	567	1316·8126	617	1455·4405	667	1595·8287	717	1737·8450
518	1182·8146	568	1319·5669	618	1458·2315	668	1598·6535	718	1740·7011
519	1185·5298	569	1322·3220	619	1461·0232	669	1601·4789	719	1743·5578
520	1188·2458	570	1325·0779	620	1463·8156	670	1604·3050	720	1746·4152
521	1190·9626	571	1327·8345	621	1466·6087	671	1607·1317	721	1749·2731
522	1193·6803	572	1330·5919	622	1469·4025	672	1609·9591	722	1752·1316
523	1196·3988	573	1333·3501	623	1472·1970	673	1612·7871	723	1754·9908
524	1199·1181	574	1336·1090	624	1474·9922	674	1615·6158	724	1757·8505
525	1201·8383	575	1338·8687	625	1477·7880	675	1618·4451	725	1760·7109
526	1204·5593	576	1341·6291	626	1480·5846	676	1621·2750	726	1763·5718
527	1207·2811	577	1344·3903	627	1483·3819	677	1624·1056	727	1766·4333
528	1210·0037	578	1347·1522	628	1486·1798	678	1626·9368	728	1769·2955
529	1212·7272	579	1349·9149	629	1488·9785	679	1629·7687	729	1772·1582
530	1215·4514	580	1352·6783	630	1491·7778	680	1632·6012	730	1775·0215
531	1218·1765	581	1355·4425	631	1494·5779	681	1635·4344	731	1777·8854
532	1220·9024	582	1358·2074	632	1497·3786	682	1638·2681	732	1780·7499
533	1223·6292	583	1360·9731	633	1500·1800	683	1641·1026	733	1783·6150
534	1226·3567	584	1363·7395	634	1502·9821	684	1643·9376	734	1786·4807
535	1229·0851	585	1366·5066	635	1505·7849	685	1646·7733	735	1789·3470
536	1231·8142	586	1369·2745	636	1508·5883	686	1649·6096	736	1792·2139
537	1234·5442	587	1372·0432	637	1511·3924	687	1652·4466	737	1795·0814
538	1237·2750	588	1374·8126	638	1514·1973	688	1655·2842	738	1797·9494
539	1240·0066	589	1377·5827	639	1517·0028	689	1658·1224	739	1800·8181
540	1242·7390	590	1380·3535	640	1519·8090	690	1660·9612	740	1803·6873
541	1245·4722	591	1383·1251	641	1522·6158	691	1663·8007	741	1806·5571
542	1248·2062	592	1385·8974	642	1525·4233	692	1666·6408	742	1809·4275
543	1250·9410	593	1388·6705	643	1528·2316	693	1669·4816	743	1812·2985
544	1253·6766	594	1391·4443	644	1531·0404	694	1672·3229	744	1815·1701
545	1256·4130	595	1394·2188	645	1533·8500	695	1675·1649	745	1818·0423
546	1259·1501	596	1396·9940	646	1536·6602	696	1678·0075	746	1820·9150
547	1261·8881	597	1399·7700	647	1539·4711	697	1680·8508	747	1823·7883
548	1264·6269	598	1402·5467	648	1542·2827	698	1683·6946	748	1826·6622
549	1267·3665	599	1405·3241	649	1545·0950	699	1686·5391	749	1829·5367
550	1270·1069	600	1408·1023	650	1547·9079	700	1689·3842	750	1832·4118

TABLE XIII. LOGARITHMS OF $n!$

n	$\log n!$	n	$\log n!$	n	$\log n!$	n	$\log n!$	n	$\log n!$
751	1835·2874	801	1979·7907	851	2125·6495	901	2272·7842	951	2421·1238
752	1838·1636	802	1982·6949	852	2128·5800	902	2275·7394	952	2424·1025
753	1841·0404	803	1985·5996	853	2131·5109	903	2278·6951	953	2427·0816
754	1843·9178	804	1988·5049	854	2134·4424	904	2281·6513	954	2430·0611
755	1846·7957	805	1991·4107	855	2137·3744	905	2284·6079	955	2433·0411
756	1849·6742	806	1994·3170	856	2140·3068	906	2287·5650	956	2436·0216
757	1852·5533	807	1997·2239	857	2143·2398	907	2290·5226	957	2439·0025
758	1855·4330	808	2000·1313	858	2146·1733	908	2293·4807	958	2441·9839
759	1858·3133	809	2003·0392	859	2149·1073	909	2296·4393	959	2444·9657
760	1861·1941	810	2005·9477	860	2152·0418	910	2299·3983	960	2447·9479
761	1864·0755	811	2008·8567	861	2154·9768	911	2302·3579	961	2450·9307
762	1866·9574	812	2011·7663	862	2157·9123	912	2305·3179	962	2453·9138
763	1869·8399	813	2014·6764	863	2160·8483	913	2308·2783	963	2456·8975
764	1872·7230	814	2017·5870	864	2163·7848	914	2311·2393	964	2459·8815
765	1875·6067	815	2020·4982	865	2166·7218	915	2314·2007	965	2462·8661
766	1878·4909	816	2023·4099	866	2169·6594	916	2317·1626	966	2465·8511
767	1881·3757	817	2026·3221	867	2172·5974	917	2320·1250	967	2468·8365
768	1884·2611	818	2029·2348	868	2175·5359	918	2323·0878	968	2471·8224
769	1887·1470	819	2032·1481	869	2178·4749	919	2326·0511	969	2474·8087
770	1890·0335	820	2035·0619	870	2181·4144	920	2329·0149	970	2477·7954
771	1892·9205	821	2037·9763	871	2184·3545	921	2331·9792	971	2480·7827
772	1895·8082	822	2040·8911	872	2187·2950	922	2334·9439	972	2483·7703
773	1898·6963	823	2043·8065	873	2190·2360	923	2337·9091	973	2486·7584
774	1901·5851	824	2046·7225	874	2193·1775	924	2340·8748	974	2489·7470
775	1904·4744	825	2049·6389	875	2196·1195	925	2343·8409	975	2492·7360
776	1907·3642	826	2052·5559	876	2199·0620	926	2346·8075	976	2495·7255
777	1910·2547	827	2055·4734	877	2202·0050	927	2349·7746	977	2498·7154
778	1913·1456	828	2058·3914	878	2204·9485	928	2352·7421	978	2501·7057
779	1916·0372	829	2061·3100	879	2207·8925	929	2355·7102	979	2504·6965
780	1918·9293	830	2064·2291	880	2210·8370	930	2358·6786	980	2507·6877
781	1921·8219	831	2067·1487	881	2213·7820	931	2361·6476	981	2510·6794
782	1924·7151	832	2070·0688	882	2216·7274	932	2364·6170	982	2513·6715
783	1927·6089	833	2072·9894	883	2219·6734	933	2367·5869	983	2516·6640
784	1930·5032	834	2075·9106	884	2222·6198	934	2370·5572	984	2519·6570
785	1933·3981	835	2078·8323	885	2225·5668	935	2373·5281	985	2522·6505
786	1936·2935	836	2081·7545	886	2228·5142	936	2376·4993	986	2525·6443
787	1939·1895	837	2084·6772	887	2231·4621	937	2379·4711	987	2528·6387
788	1942·0860	838	2087·6005	888	2234·4106	938	2382·4433	988	2531·6334
789	1944·9831	839	2090·5242	889	2237·3595	939	2385·4159	989	2534·6286
790	1947·8807	840	2093·4485	890	2240·3088	940	2388·3891	990	2537·6242
791	1950·7789	841	2096·3733	891	2243·2587	941	2391·3627	991	2540·6203
792	1953·6776	842	2099·2986	892	2246·2091	942	2394·3367	992	2543·6168
793	1956·5769	843	2102·2244	893	2249·1599	943	2397·3112	993	2546·6138
794	1959·4767	844	2105·1508	894	2252·1113	944	2400·2862	994	2549·6112
795	1962·3771	845	2108·0776	895	2255·0631	945	2403·2616	995	2552·6090
796	1965·2780	846	2111·0050	896	2258·0154	946	2406·2375	996	2555·6073
797	1968·1794	847	2113·9329	897	2260·9682	947	2409·2139	997	2558·6059
798	1971·0814	848	2116·8613	898	2263·9215	948	2412·1907	998	2561·6051
799	1973·9840	849	2119·7902	899	2266·8752	949	2415·1679	999	2564·6046
800	1976·8871	850	2122·7196	900	2269·8295	950	2418·1457	1000	2567·6046

$$\log n! = (n + \tfrac{1}{2}) \log n - \left(n - \frac{1}{12n}\right) \log e + \log \sqrt{2\pi} + R, \text{ where } -1/360\, n^3 < R < 0.$$

TABLE XIV. LOGARITHMS OF BINOMIAL COEFFICIENTS. $\log \binom{n}{x}$.

x	—	$n=2$	$n=3$	$n=4$	$n=5$	$n=6$	$n=7$	$n=8$	$n=9$	$n=10$
1		0·3010	0·4771	0·6021	0·6990	0·7782	0·8451	0·9031	0·9542	1·0000
2			0·4771	0·7782	1·0000	1·1761	1·3222	1·4472	1·5563	1·6532
3					1·0000	1·3010	1·5441	1·7482	1·9243	2·0792
4							1·5441	1·8451	2·1004	2·3222
5									2·1004	2·4014

x	$n=11$	$n=12$	$n=13$	$n=14$	$n=15$	$n=16$	$n=17$	$n=18$	$n=19$	$n=20$
1	1·0414	1·0792	1·1139	1·1461	1·1761	1·2041	1·2304	1·2553	1·2788	1·3010
2	1·7404	1·8195	1·8921	1·9590	2·0212	2·0792	2·1335	2·1847	2·2330	2·2788
3	2·2175	2·3424	2·4564	2·5611	2·6580	2·7482	2·8325	2·9117	2·9863	3·0569
4	2·5185	2·6946	2·8543	3·0004	3·1351	3·2601	3·3766	3·4857	3·5884	3·6853
5	2·6646	2·8987	3·1096	3·3015	3·4776	3·6402	3·7916	3·9329	4·0655	4·1904
6	2·6646	2·9657	3·2345	3·4776	3·6994	3·9035	4·0926	4·2687	4·4335	4·5884
7			3·2345	3·5355	3·8085	4·0584	4·2889	4·5028	4·7023	4·8894
8					3·8085	4·1096	4·3858	4·6411	4·8784	5·1003
9							4·3858	4·6868	4·9656	5·2252
10									4·9656	5·2666

x	$n=21$	$n=22$	$n=23$	$n=24$	$n=25$	$n=26$	$n=27$	$n=28$	$n=29$	$n=30$
1	1·3222	1·3424	1·3617	1·3802	1·3979	1·4150	1·4314	1·4472	1·4624	1·4771
2	2·3222	2·3636	2·4031	2·4409	2·4771	2·5119	2·5453	2·5775	2·6085	2·6385
3	3·1239	3·1875	3·2482	3·3062	3·3617	3·4150	3·4661	3·5153	3·5628	3·6085
4	3·7771	3·8642	3·9472	4·0264	4·1021	4·1746	4·2443	4·3112	4·3757	4·4378
5	4·3085	4·4205	4·5270	4·6284	4·7253	4·8181	4·9070	4·9925	5·0747	5·1538
6	4·7345	4·8728	5·0041	5·1290	5·2482	5·3622	5·4713	5·5760	5·6767	5·7736
7	5·0655	5·2318	5·3894	5·5392	5·6819	5·8181	5·9484	6·0734	6·1933	6·3087
8	5·3085	5·5048	5·6905	5·8666	6·0341	6·1938	6·3464	6·4925	6·6327	6·7674
9	5·4682	5·6967	5·9123	6·1164	6·3103	6·4948	6·6709	6·8393	7·0007	7·1556
10	5·5474	5·8107	6·0585	6·2925	6·5144	6·7252	6·9262	7·1180	7·3017	7·4778
11	5·5474	5·8485	6·1310	6·3973	6·6491	6·8880	7·1152	7·3319	7·5390	7·7374
12			6·1310	6·4320	6·7160	6·9849	7·2401	7·4832	7·7151	7·9370
13					6·7160	7·0171	7·3023	7·5734	7·8316	8·0783
14							7·3023	7·6033	7·8896	8·1626
15									7·8896	8·1907

x	$n=31$	$n=32$	$n=33$	$n=34$	$n=35$	$n=36$	$n=37$	$n=38$	$n=39$	$n=40$
1	1·4914	1·5052	1·5185	1·5315	1·5441	1·5563	1·5682	1·5798	1·5911	1·6021
2	2·6675	2·6955	2·7226	2·7490	2·7745	2·7993	2·8235	2·8470	2·8698	2·8921
3	3·6527	3·6955	3·7369	3·7770	3·8159	3·8537	3·8904	3·9261	3·9609	3·9948
4	4·4978	4·5558	4·6119	4·6663	4·7190	4·7702	4·8198	4·8681	4·9151	4·9609
5	5·2302	5·3040	5·3754	5·4444	5·5114	5·5763	5·6394	5·7007	5·7602	5·8182
6	5·8670	5·9572	6·0444	6·1287	6·2104	6·2895	6·3664	6·4410	6·5136	6·5841
7	6·4199	6·5271	6·6306	6·7308	6·8277	6·9216	7·0126	7·1011	7·1870	7·2705
8	6·8970	7·0219	7·1425	7·2590	7·3717	7·4809	7·5867	7·6893	7·7890	7·8860
9	7·3045	7·4479	7·5862	7·7198	7·8489	7·9738	8·0948	8·2122	8·3262	8·4369
10	7·6469	7·8096	7·9664	8·1177	8·2638	8·4052	8·5420	8·6746	8·8033	8·9282
11	7·9277	8·1107	8·2868	8·4565	8·6204	8·7787	8·9320	9·0804	9·2243	9·3640
12	8·1496	8·3537	8·5500	8·7391	8·9214	9·0975	9·2678	9·4326	9·5923	9·7472
13	8·3144	8·5408	8·7583	8·9675	9·1692	9·3638	9·5518	9·7336	9·9097	10·0804
14	8·4235	8·6734	8·9132	9·1436	9·3655	9·5794	9·7858	9·9854	10·1785	10·3656
15	8·4779	8·7526	9·0158	9·2686	9·5116	9·7457	9·9715	10·1895	10·4004	10·6045

TABLE XIV. LOGARITHMS OF BINOMIAL COEFFICIENTS. $\log \binom{n}{x}$. **77**

x	n=31	n=32	n=33	n=34	n=35	n=36	n=37	n=38	n=39	n=40
16	8.4779	8.7789	9.0670	9.3432	9.6085	9.8638	10.1098	10.3471	10.5765	10.7983
17			9.0670	9.3680	9.6568	9.9344	10.2015	10.4591	10.7078	10.9481
18					9.6568	9.9579	10.2473	10.5261	10.7949	11.0545
19							10.2473	10.5483	10.8384	11.1182
20									10.8384	11.1394

x	n=41	n=42	n=43	n=44	n=45	n=46	n=47	n=48	n=49	n=50
1	1.6128	1.6232	1.6335	1.6435	1.6532	1.6628	1.6721	1.6812	1.6902	1.6990
2	2.9138	2.9350	2.9557	2.9759	2.9956	3.0149	3.0338	3.0523	3.0704	3.0881
3	4.0278	4.0599	4.0914	4.1220	4.1520	4.1813	4.2099	4.2379	4.2654	4.2923
4	5.0055	5.0489	5.0914	5.1327	5.1732	5.2127	5.2513	5.2891	5.3261	5.3623
5	5.8747	5.9298	5.9834	6.0358	6.0870	6.1370	6.1858	6.2336	6.2803	6.3261
6	6.6529	6.7198	6.7851	6.8487	6.9109	6.9716	7.0309	7.0889	7.1456	7.2011
7	7.3518	7.4310	7.5082	7.5834	7.6569	7.7286	7.7986	7.8670	7.9340	7.9995
8	7.9802	8.0720	8.1614	8.2485	8.3336	8.4165	8.4976	8.5767	8.6542	8.7299
9	8.5445	8.6492	8.7512	8.8506	8.9475	9.0421	9.1344	9.2246	9.3127	9.3989
10	9.0496	9.1677	9.2827	9.3947	9.5038	9.6103	9.7142	9.8156	9.9148	10.0117
11	9.4996	9.6315	9.7598	9.8848	10.0065	10.1252	10.2410	10.3540	10.4644	10.5723
12	9.8976	10.0437	10.1858	10.3241	10.4588	10.5901	10.7181	10.8430	10.9650	11.0842
13	10.2460	10.4069	10.5632	10.7153	10.8634	11.0076	11.1482	11.2854	11.4193	11.5501
14	10.5470	10.7231	10.8942	11.0605	11.2224	11.3800	11.5336	11.6833	11.8295	11.9721
15	10.8023	10.9942	11.1805	11.3616	11.5377	11.7090	11.8760	12.0387	12.1974	12.3523
16	11.0132	11.2214	11.4235	11.6198	11.8107	11.9963	12.1770	12.3531	12.5248	12.6923
17	11.1807	11.4060	11.6245	11.8365	12.0426	12.2430	12.4379	12.6278	12.8129	12.9933
18	11.3056	11.5486	11.7842	12.0126	12.2345	12.4501	12.6598	12.8639	13.0627	13.2566
19	11.3886	11.6501	11.9033	12.1489	12.3871	12.6185	12.8434	13.0623	13.2754	13.4830
20	11.4300	11.7108	11.9825	12.2458	12.5010	12.7488	12.9896	13.2236	13.4514	13.6733
21	11.4300	11.7310	12.0220	12.3038	12.5768	12.8416	13.0987	13.3486	13.5916	13.8282
22			12.0220	12.3231	12.6146	12.8971	13.1713	13.4375	13.6964	13.9482
23					12.6146	12.9156	13.2075	13.4908	13.7660	14.0336
24							13.2075	13.5085	13.8008	14.0848
25									13.8008	14.1018

x	n=51	n=52	n=53	n=54	n=55	n=56	n=57	n=58	n=59	n=60
1	1.7076	1.7160	1.7243	1.7324	1.7404	1.7482	1.7559	1.7634	1.7709	1.7782
2	3.1055	3.1225	3.1392	3.1556	3.1717	3.1875	3.2030	3.2183	3.2333	3.2480
3	4.3186	4.3444	4.3697	4.3945	4.4189	4.4428	4.4663	4.4893	4.5120	4.5343
4	5.3978	5.4325	5.4666	5.5000	5.5328	5.5650	5.5966	5.6276	5.6581	5.6881
5	6.3709	6.4148	6.4578	6.5000	6.5414	6.5820	6.6219	6.6611	6.6995	6.7373
6	7.2555	7.3087	7.3609	7.4121	7.4622	7.5115	7.5598	7.6072	7.6538	7.6995
7	8.0636	8.1264	8.1879	8.2482	8.3073	8.3653	8.4222	8.4781	8.5329	8.5868
8	8.8040	8.8765	8.9476	9.0172	9.0855	9.1524	9.2181	9.2826	9.3459	9.4080
9	9.4832	9.5657	9.6466	9.7257	9.8033	9.8794	9.9541	10.0273	10.0992	10.1698
10	10.1065	10.1992	10.2900	10.3790	10.4661	10.5515	10.6353	10.7175	10.7982	10.8773
11	10.6778	10.7811	10.8821	10.9810	11.0779	11.1729	11.2660	11.3573	11.4470	11.5349
12	11.2007	11.3147	11.4262	11.5353	11.6422	11.7469	11.8496	11.9503	12.0490	12.1459
13	11.6778	11.8028	11.9250	12.0446	12.1617	12.2764	12.3889	12.4991	12.6072	12.7132
14	12.1115	12.2477	12.3809	12.5113	12.6388	12.7638	12.8862	13.0062	13.1238	13.2392
15	12.5036	12.6514	12.7959	12.9372	13.0755	13.2109	13.3436	13.4735	13.6009	13.7259

TABLE XIV. LOGARITHMS OF BINOMIAL COEFFICIENTS. $\log \binom{n}{x}$.

x	$n=51$	$n=52$	$n=53$	$n=54$	$n=55$	$n=56$	$n=57$	$n=58$	$n=59$	$n=60$
16	12.8558	13.0155	13.1716	13.3242	13.4735	13.6196	13.7627	13.9029	14.0403	14.1750
17	13.1694	13.3413	13.5093	13.6735	13.8341	13.9912	14.1450	14.2957	14.4433	14.5880
18	13.4456	13.6301	13.8104	13.9864	14.1586	14.3270	14.4918	14.6532	14.8113	14.9662
19	13.6854	13.8829	14.0757	14.2640	14.4481	14.6280	14.8041	14.9765	15.1453	15.3107
20	13.8895	14.1004	14.3061	14.5070	14.7033	14.8952	15.0829	15.2665	15.4463	15.6224
21	14.0586	14.2833	14.5024	14.7163	14.9252	15.1293	15.3289	15.5241	15.7152	15.9022
22	14.1933	14.4322	14.6651	14.8924	15.1142	15.3309	15.5427	15.7499	15.9525	16.1509
23	14.2940	14.5476	14.7948	15.0358	15.2710	15.5007	15.7251	15.9444	16.1590	16.3689
24	14.3610	14.6298	14.8917	15.1470	15.3960	15.6390	15.8764	16.1083	16.3351	16.5569
25	14.3944	14.6790	14.9561	15.2261	15.4894	15.7462	15.9969	16.2418	16.4812	16.7153
26	14.3944	14.6954	14.9883	15.2736	15.5515	15.8226	16.0871	16.3454	16.5977	16.8444
27			14.9883	15.2894	15.5826	15.8683	16.1471	16.4192	16.6849	16.9445
28					15.5826	15.8836	16.1771	16.4634	16.7429	17.0159
29							16.1771	16.4781	16.7718	17.0586
30									16.7718	17.0729

x	$n=61$	$n=62$	$n=63$	$n=64$	$n=65$	$n=66$	$n=67$	$n=68$	$n=69$	$n=70$
1	1.7853	1.7924	1.7993	1.8062	1.8129	1.8195	1.8261	1.8325	1.8388	1.8451
2	3.2625	3.2767	3.2907	3.3045	3.3181	3.3314	3.3446	3.3576	3.3703	3.3829
3	4.5562	4.5777	4.5989	4.6198	4.6403	4.6605	4.6804	4.7000	4.7193	4.7383
4	5.7176	5.7465	5.7750	5.8030	5.8306	5.8578	5.8845	5.9108	5.9368	5.9623
5	6.7745	6.8110	6.8469	6.8822	6.9170	6.9512	6.9849	7.0180	7.0507	7.0829
6	7.7445	7.7887	7.8322	7.8749	7.9170	7.9584	7.9991	8.0392	8.0787	8.1177
7	8.6398	8.6918	8.7429	8.7932	8.8427	8.8914	8.9393	8.9865	9.0330	9.0787
8	9.4691	9.5291	9.5880	9.6460	9.7031	9.7592	9.8144	9.8688	9.9223	9.9750
9	10.2391	10.3072	10.3742	10.4400	10.5047	10.5684	10.6310	10.6927	10.7534	10.8131
10	10.9551	11.0315	11.1066	11.1803	11.2529	11.3242	11.3944	11.4635	11.5315	11.5985
11	11.6213	11.7061	11.7894	11.8713	11.9519	12.0310	12.1089	12.1856	12.2610	12.3352
12	12.2411	12.3345	12.4263	12.5164	12.6051	12.6922	12.7779	12.8623	12.9452	13.0269
13	12.8173	12.9195	13.0199	13.1185	13.2154	13.3107	13.4044	13.4965	13.5872	13.6764
14	13.3524	13.4636	13.5727	13.6799	13.7853	13.8888	13.9906	14.0907	14.1892	14.2861
15	13.8484	13.9687	14.0868	14.2028	14.3168	14.4287	14.5388	14.6470	14.7535	14.8582
16	14.3071	14.4367	14.5640	14.6889	14.8116	14.9322	15.0507	15.1672	15.2818	15.3945
17	14.7298	14.8690	15.0056	15.1397	15.2714	15.4007	15.5278	15.6527	15.7756	15.8964
18	15.1180	15.2670	15.4131	15.5565	15.6973	15.8356	15.9715	16.1050	16.2363	16.3654
19	15.4727	15.6317	15.7875	15.9405	16.0907	16.2381	16.3829	16.5253	16.6651	16.8027
20	15.7950	15.9641	16.1300	16.2927	16.4524	16.6092	16.7632	16.9144	17.0631	17.2092
21	16.0855	16.2651	16.4412	16.6139	16.7834	16.9497	17.1130	17.2734	17.4311	17.5860
22	16.3452	16.5355	16.7220	16.9050	17.0844	17.2605	17.4334	17.6031	17.7699	17.9337
23	16.5745	16.7758	16.9731	17.1665	17.3562	17.5422	17.7249	17.9042	18.0802	18.2532
24	16.7741	16.9867	17.1949	17.3991	17.5992	17.7955	17.9881	18.1772	18.3628	18.5451
25	16.9443	17.1685	17.3881	17.6032	17.8140	18.0208	18.2236	18.4227	18.6181	18.8099
26	17.0857	17.3217	17.5529	17.7793	18.0011	18.2186	18.4319	18.6412	18.8465	19.0482
27	17.1984	17.4467	17.6897	17.9277	18.1608	18.3893	18.6133	18.8330	19.0486	19.2603
28	17.2827	17.5436	17.7989	18.0487	18.2935	18.5332	18.7682	18.9987	19.2247	19.4466
29	17.3388	17.6127	17.8805	18.1426	18.3993	18.6506	18.8969	19.1383	19.3751	19.6074
30	17.3662	17.6541	17.9349	18.2096	18.4784	18.7417	18.9996	19.2523	19.5001	19.7431
31	17.3662	17.6679	17.9620	18.2497	18.5311	18.8066	19.0764	19.3407	19.5998	19.8538
32			17.9620	18.2631	18.5575	18.8455	19.1275	19.4038	19.6744	19.9397
33					18.5575	18.8585	19.1531	19.4415	19.7241	20.0010
34							19.1531	19.4541	19.7489	20.0377
35									19.7489	20.0499

TABLE XIV. LOGARITHMS OF BINOMIAL COEFFICIENTS. $\log \binom{n}{x}$.

x	n = 71	n = 72	n = 73	n = 74	n = 75	n = 76	n = 77	n = 78	n = 79	n = 80
1	1·8513	1·8573	1·8633	1·8692	1·8751	1·8808	1·8865	1·8921	1·8976	1·9031
2	3·3953	3·4076	3·4196	3·4315	3·4433	3·4548	3·4663	3·4776	3·4887	3·4997
3	4·7571	4·7755	4·7938	4·8117	4·8295	4·8470	4·8642	4·8812	4·8981	4·9147
4	5·9875	6·0123	6·0368	6·0609	6·0847	6·1082	6·1314	6·1542	6·1768	6·1991
5	7·1146	7·1459	7·1767	7·2071	7·2370	7·2666	7·2957	7·3245	7·3529	7·3809
6	8·1560	8·1938	8·2310	8·2678	8·3040	8·3397	8·3749	8·4097	8·4440	8·4778
7	9·1238	9·1682	9·2120	9·2552	9·2977	9·3397	9·3811	9·4219	9·4622	9·5020
8	10·0269	10·0786	10·1285	10·1782	10·2271	10·2754	10·3231	10·3701	10·4165	10·4622
9	10·8720	10·9300	10·9871	11·0435	11·0990	11·1537	11·2077	11·2609	11·3135	11·3653
10	11·6644	11·7293	11·7933	11·8564	11·9185	11·9798	12·0402	12·0998	12·1586	12·2166
11	12·4083	12·4803	12·5513	12·6212	12·6900	12·7579	12·8249	12·8909	12·9560	13·0203
12	13·1073	13·1865	13·2645	13·3413	13·4170	13·4917	13·5652	13·6378	13·7094	13·7799
13	13·7642	13·8507	13·9359	14·0198	14·1024	14·1839	14·2642	14·3434	14·4215	14·4985
14	14·3815	14·4754	14·5679	14·6590	14·7487	14·8371	14·9243	15·0102	15·0949	15·1784
15	14·9613	15·0628	15·1626	15·2610	15·3579	15·4534	15·5475	15·6403	15·7317	15·8219
16	15·5054	15·6145	15·7220	15·8278	15·9320	16·0346	16·1358	16·2355	16·3338	16·4307
17	16·0153	16·1322	16·2474	16·3607	16·4724	16·5823	16·6907	16·7974	16·9027	17·0064
18	16·4924	16·6173	16·7403	16·8613	16·9805	17·0979	17·2136	17·3275	17·4398	17·5505
19	16·9379	17·0710	17·2019	17·3308	17·4576	17·5826	17·7057	17·8269	17·9464	18·0641
20	17·3529	17·4942	17·6333	17·7701	17·9048	18·0374	18·1680	18·2967	18·4235	18·5484
21	17·7382	17·8880	18·0353	18·1803	18·3230	18·4634	18·6017	18·7379	18·8721	19·0044
22	18·0948	18·2532	18·4089	18·5621	18·7129	18·8613	19·0075	19·1514	19·2931	19·4328
23	18·4233	18·5904	18·7547	18·9164	19·0755	19·2320	19·3861	19·5378	19·6873	19·8345
24	18·7243	18·9004	19·0735	19·2438	19·4113	19·5761	19·7383	19·8980	20·0553	20·2102
25	18·9984	19·1837	19·3658	19·5448	19·7209	19·8941	20·0646	20·2324	20·3977	20·5604
26	19·2462	19·4408	19·6320	19·8200	20·0049	20·1867	20·3657	20·5417	20·7151	20·8858
27	19·4681	19·6722	19·8728	20·0699	20·2637	20·4543	20·6419	20·8264	21·0080	21·1868
28	19·6644	19·8783	20·0884	20·2948	20·4978	20·6974	20·8937	21·0868	21·2769	21·4639
29	19·8354	20·0593	20·2792	20·4952	20·7075	20·9162	21·1215	21·3234	21·5220	21·7175
30	19·9816	20·2157	20·4455	20·6713	20·8931	21·1112	21·3256	21·5364	21·7439	21·9480
31	20·1030	20·3475	20·5876	20·8234	21·0550	21·2826	21·5063	21·7263	21·9427	22·1556
32	20·1999	20·4552	20·7057	20·9517	21·1933	21·4307	21·6639	21·8933	22·1188	22·3406
33	20·2725	20·5387	20·8000	21·0564	21·3082	21·5556	21·7986	22·0375	22·2724	22·5034
34	20·3208	20·5983	20·8706	21·1377	21·4000	21·6576	21·9106	22·1592	22·4037	22·6440
35	20·3449	20·6340	20·9176	21·1957	21·4687	21·7368	22·0000	22·2586	22·5128	22·7627
36	20·3449	20·6459	20·9410	21·2305	21·5145	21·7932	22·0670	22·3358	22·6000	22·8596
37			20·9410	21·2421	21·5374	21·8271	22·1115	22·3908	22·6652	22·9348
38				21·5374	21·8384	22·1338	22·4238	22·7087	22·9885
39							22·1338	22·4348	22·7304	23·0207
40									22·7304	23·0314

TABLE XIV. LOGARITHMS OF BINOMIAL COEFFICIENTS. $\log \binom{n}{x}$.

x	n = 81	n = 82	n = 83	n = 84	n = 85	n = 86	n = 87	n = 88	n = 89	n = 90
1	1·9085	1·9138	1·9191	1·9243	1·9294	1·9345	1·9395	1·9445	1·9494	1·9542
2	3·5105	3·5213	3·5319	3·5423	3·5527	3·5629	3·5730	3·5830	3·5928	3·6026
3	4·9311	4·9472	4·9632	4·9790	4·9946	5·0100	5·0253	5·0403	5·0552	5·0700
4	6·2211	6·2428	6·2643	6·2854	6·3064	6·3271	6·3475	6·3677	6·3877	6·4074
5	7·4086	7·4359	7·4629	7·4896	7·5159	7·5419	7·5676	7·5930	7·6181	7·6430
6	8·5113	8·5445	8·5769	8·6090	8·6408	8·6722	8·7033	8·7339	8·7643	8·7942
7	9·5412	9·5800	9·6182	9·6560	9·6934	9·7302	9·7667	9·8027	9·8382	9·8734
8	10·5074	10·5520	10·5960	10·6394	10·6824	10·7248	10·7667	10·8081	10·8490	10·8894
9	11·4165	11·4669	11·5168	11·5660	11·6146	11·6626	11·7100	11·7569	11·8032	11·8490
10	12·2738	12·3303	12·3860	12·4411	12·4954	12·5491	12·6021	12·6545	12·7063	12·7574
11	13·0837	13·1462	13·2080	13·2689	13·3291	13·3885	13·4472	13·5052	13·5625	13·6191
12	13·8496	13·9183	13·9861	14·0531	14·1191	14·1844	14·2489	14·3125	14·3754	14·4376
13	14·5745	14·6494	14·7234	14·7964	14·8685	14·9397	15·0100	15·0794	15·1480	15·2157
14	15·2609	15·3422	15·4224	15·5016	15·5797	15·6569	15·7331	15·8083	15·8827	15·9561
15	15·9108	15·9986	16·0851	16·1706	16·2549	16·3381	16·4203	16·5015	16·5816	16·6608
16	16·5263	16·6205	16·7135	16·8053	16·8959	16·9853	17·0735	17·1607	17·2468	17·3318
17	17·1087	17·2096	17·3092	17·4074	17·5043	17·5999	17·6943	17·7876	17·8796	17·9705
18	17·6596	17·7673	17·8734	17·9782	18·0815	18·1835	18·2842	18·3836	18·4817	18·5786
19	18·1802	18·2947	18·4076	18·5190	18·6288	18·7373	18·8443	18·9499	19·0542	19·1572
20	18·6716	18·7930	18·9127	19·0308	19·1473	19·2623	19·3757	19·4877	19·5983	19·7074
21	19·1347	19·2632	19·3899	19·5148	19·6380	19·7596	19·8796	19·9980	20·1149	20·2303
22	19·5704	19·7061	19·8398	19·9717	20·1018	20·2301	20·3567	20·4817	20·6050	20·7267
23	19·9795	20·1225	20·2634	20·4024	20·5394	20·6746	20·8079	20·9395	21·0693	21·1975
24	20·3628	20·5131	20·6614	20·8075	20·9516	21·0937	21·2339	21·3722	21·5087	21·6434
25	20·7207	20·8786	21·0343	21·1877	21·3390	21·4882	21·6353	21·7804	21·9236	22·0650
26	21·0539	21·2195	21·3827	21·5436	21·7022	21·8585	22·0127	22·1648	22·3148	22·4629
27	21·3629	21·5364	21·7073	21·8757	22·0416	22·2053	22·3667	22·5258	22·6828	22·8377
28	21·6481	21·8296	22·0083	22·1844	22·3579	22·5290	22·6977	22·8640	23·0280	23·1899
29	21·9100	22·0996	22·2862	22·4702	22·6514	22·8300	23·0061	23·1797	23·3510	23·5199
30	22·1489	22·3467	22·5415	22·7334	22·9225	23·1088	23·2924	23·4735	23·6520	23·8281
31	22·3651	22·5714	22·7744	22·9744	23·1715	23·3656	23·5569	23·7455	23·9315	24·1149
32	22·5589	22·7738	22·9853	23·1936	23·3987	23·6008	23·8000	23·9963	24·1898	24·3806
33	22·7306	22·9542	23·1743	23·3911	23·6045	23·8147	24·0218	24·2259	24·4271	24·6255
34	22·8804	23·1130	23·3418	23·5671	23·7890	24·0075	24·2227	24·4348	24·6438	24·8499
35	23·0084	23·2501	23·4880	23·7220	23·9525	24·1794	24·4029	24·6231	24·8401	25·0540
36	23·1149	23·3659	23·6129	23·8559	24·0952	24·3307	24·5626	24·7911	25·0162	25·2381
37	23·1999	23·4605	23·7168	23·9690	24·2172	24·4615	24·7020	24·9389	25·1723	25·4023
38	23·2635	23·5339	23·7998	24·0613	24·3186	24·5719	24·8212	25·0667	25·3085	25·5468
39	23·3059	23·5863	23·8619	24·1330	24·3996	24·6620	24·9203	25·1746	25·4250	25·6717
40	23·3271	23·6177	23·9033	24·1841	24·4603	24·7321	24·9995	25·2627	25·5219	25·7772
41	23·3271	23·6282	23·9240	24·2148	24·5008	24·7821	25·0588	25·3312	25·5994	25·8634
42		23·6282	23·9240	24·2250	24·5210	24·8120	25·0983	25·3801	25·6573	25·9303
43			23·9240	24·2250	24·5210	24·8220	25·1181	25·4093	25·6960	25·9781
44							25·1181	25·4191	25·7153	26·0068
45									25·7153	26·0163

TABLE XIV. LOGARITHMS OF BINOMIAL COEFFICIENTS. $\log\binom{n}{x}$.

x	n = 91	n = 92	n = 93	n = 94	n = 95	n = 96	n = 97	n = 98	n = 99	n = 100
1	1.9590	1.9638	1.9685	1.9731	1.9777	1.9823	1.9868	1.9912	1.9956	2.0000
2	3.6123	3.6218	3.6312	3.6406	3.6498	3.6590	3.6680	3.6770	3.6858	3.6946
3	5.0845	5.0989	5.1132	5.1272	5.1412	5.1550	5.1686	5.1821	5.1955	5.2087
4	6.4269	6.4463	6.4653	6.4842	6.5029	6.5214	6.5397	6.5578	6.5757	6.5934
5	7.6675	7.6918	7.7158	7.7395	7.7630	7.7862	7.8092	7.8319	7.8544	7.8767
6	8.8238	8.8531	8.8821	8.9107	8.9391	8.9671	8.9948	9.0223	9.0494	9.0763
7	9.9082	9.9425	9.9765	10.0101	10.0434	10.0762	10.1088	10.1410	10.1728	10.2043
8	10.9294	10.9689	11.0079	11.0466	11.0848	11.1225	11.1599	11.1969	11.2335	11.2697
9	11.8942	11.9389	11.9831	12.0268	12.0700	12.1128	12.1551	12.1969	12.2383	12.2793
10	12.8080	12.8580	12.9074	12.9562	13.0045	13.0523	13.0996	13.1463	13.1925	13.2383
11	13.6751	13.7304	13.7851	13.8391	13.8926	13.9454	13.9977	14.0494	14.1005	14.1512
12	14.4990	14.5597	14.6197	14.6790	14.7377	14.7957	14.8530	14.9097	14.9658	15.0214
13	15.2827	15.3488	15.4142	15.4789	15.5428	15.6060	15.6685	15.7303	15.7914	15.8519
14	16.0287	16.1003	16.1712	16.2412	16.3105	16.3789	16.4466	16.5136	16.5798	16.6453
15	16.7391	16.8163	16.8927	16.9682	17.0429	17.1167	17.1896	17.2618	17.3331	17.4037
16	17.4157	17.4987	17.5807	17.6617	17.7418	17.8210	17.8993	17.9767	18.0533	18.1290
17	18.0604	18.1491	18.2368	18.3234	18.4090	18.4937	18.5773	18.6601	18.7419	18.8228
18	18.6743	18.7689	18.8623	18.9546	19.0458	19.1360	19.2252	19.3133	19.4005	19.4866
19	19.2589	19.3594	19.4586	19.5567	19.6536	19.7494	19.8440	19.9376	20.0302	20.1217
20	19.8152	19.9216	20.0268	20.1307	20.2334	20.3348	20.4351	20.5342	20.6322	20.7292
21	20.3442	20.4568	20.5679	20.6777	20.7862	20.8934	20.9994	21.1041	21.2076	21.3100
22	20.8469	20.9656	21.0828	21.1986	21.3130	21.4261	21.5378	21.6482	21.7573	21.8652
23	21.3240	21.4490	21.5723	21.6942	21.8146	21.9336	22.0511	22.1673	22.2821	22.3956
24	21.7763	21.9076	22.0372	22.1653	22.2917	22.4167	22.5401	22.6621	22.7827	22.9019
25	22.2045	22.3422	22.4781	22.6124	22.7450	22.8761	23.0055	23.1334	23.2598	23.3847
26	22.6090	22.7533	22.8957	23.0363	23.1752	23.3123	23.4479	23.5818	23.7141	23.8448
27	22.9906	23.1415	23.2904	23.4374	23.5827	23.7261	23.8678	24.0077	24.1460	24.2827
28	23.3496	23.5072	23.6628	23.8164	23.9680	24.1178	24.2657	24.4118	24.5562	24.6989
29	23.6865	23.8510	24.0133	24.1735	24.3317	24.4879	24.6421	24.7945	24.9451	25.0938
30	24.0018	24.1732	24.3424	24.5093	24.6741	24.8368	24.9975	25.1563	25.3130	25.4679
31	24.2958	24.4742	24.6503	24.8241	24.9957	25.1650	25.3322	25.4974	25.6605	25.8217
32	24.5688	24.7544	24.9376	25.1183	25.2967	25.4728	25.6466	25.8183	25.9879	26.1554
33	24.8211	25.0141	25.2044	25.3922	25.5775	25.7604	25.9410	26.1194	26.2954	26.4694
34	25.0530	25.2534	25.4511	25.6460	25.8384	26.0283	26.2157	26.4008	26.5835	26.7640
35	25.2649	25.4728	25.6778	25.8801	26.0797	26.2766	26.4710	26.6629	26.8524	27.0394
36	25.4568	25.6724	25.8850	26.0947	26.3015	26.5057	26.7071	26.9059	27.1022	27.2961
37	25.6289	25.8523	26.0726	26.2899	26.5042	26.7156	26.9242	27.1301	27.3334	27.5340
38	25.7815	26.0130	26.2410	26.4660	26.6878	26.9067	27.1226	27.3357	27.5460	27.7536
39	25.9147	26.1543	26.3903	26.6231	26.8526	27.0790	27.3024	27.5228	27.7402	27.9549
40	26.0287	26.2765	26.5207	26.7614	26.9988	27.2329	27.4638	27.6916	27.9163	28.1382
41	26.1235	26.3797	26.6322	26.8810	27.1264	27.3683	27.6068	27.8422	28.0744	28.3036
42	26.1992	26.4640	26.7249	26.9820	27.2355	27.4854	27.7318	27.9748	28.2146	28.4512
43	26.2559	26.5295	26.7990	27.0646	27.3263	27.5843	27.8387	28.0895	28.3370	28.5811
44	26.2937	26.5763	26.8545	27.1287	27.3989	27.6651	27.9276	28.1865	28.4417	28.6935
45	26.3126	26.6043	26.8915	27.1745	27.4532	27.7279	27.9987	28.2656	28.5289	28.7885
46	26.3126	26.6136	26.9100	27.2019	27.4894	27.7727	28.0519	28.3272	28.5985	28.8661
47			26.9100	27.2110	27.5075	27.7996	28.0874	28.3711	28.6507	28.9264
48					27.5075	27.8086	28.1051	28.3974	28.6855	28.9694
49							28.1051	28.4062	28.7028	28.9953
50									28.7028	29.0039

TABLE XV. SQUARES

	0	1	2	3	4	5	6	7	8	9
0	0	1	4	9	16	25	36	49	64	81
10	100	121	144	169	196	225	256	289	324	361
20	400	441	484	529	576	625	676	729	784	841
30	900	961	1024	1089	1156	1225	1296	1369	1444	1521
40	1600	1681	1764	1849	1936	2025	2116	2209	2304	2401
50	2500	2601	2704	2809	2916	3025	3136	3249	3364	3481
60	3600	3721	3844	3969	4096	4225	4356	4489	4624	4761
70	4900	5041	5184	5329	5476	5625	5776	5929	6084	6241
80	6400	6561	6724	6889	7056	7225	7396	7569	7744	7921
90	8100	8281	8464	8649	8836	9025	9216	9409	9604	9801
100	10000	10201	10404	10609	10816	11025	11236	11449	11664	11881
110	12100	12321	12544	12769	12996	13225	13456	13689	13924	14161
120	14400	14641	14884	15129	15376	15625	15876	16129	16384	16641
130	16900	17161	17424	17689	17956	18225	18496	18769	19044	19321
140	19600	19881	20164	20449	20736	21025	21316	21609	21904	22201
150	22500	22801	23104	23409	23716	24025	24336	24649	24964	25281
160	25600	25921	26244	26569	26896	27225	27556	27889	28224	28561
170	28900	29241	29584	29929	30276	30625	30976	31329	31684	32041
180	32400	32761	33124	33489	33856	34225	34596	34969	35344	35721
190	36100	36481	36864	37249	37636	38025	38416	38809	39204	39601
200	40000	40401	40804	41209	41616	42025	42436	42849	43264	43681
210	44100	44521	44944	45369	45796	46225	46656	47089	47524	47961
220	48400	48841	49284	49729	50176	50625	51076	51529	51984	52441
230	52900	53361	53824	54289	54756	55225	55696	56169	56644	57121
240	57600	58081	58564	59049	59536	60025	60516	61009	61504	62001
250	62500	63001	63504	64009	64516	65025	65536	66049	66564	67081
260	67600	68121	68644	69169	69696	70225	70756	71289	71824	72361
270	72900	73441	73984	74529	75076	75625	76176	76729	77284	77841
280	78400	78961	79524	80089	80656	81225	81796	82369	82944	83521
290	84100	84681	85264	85849	86436	87025	87616	88209	88804	89401
300	90000	90601	91204	91809	92416	93025	93636	94249	94864	95481
310	96100	96721	97344	97969	98596	99225	99856	100489	101124	101761
320	102400	103041	103684	104329	104976	105625	106276	106929	107584	108241
330	108900	109561	110224	110889	111556	112225	112896	113569	114244	114921
340	115600	116281	116964	117649	118336	119025	119716	120409	121104	121801
350	122500	123201	123904	124609	125316	126025	126736	127449	128164	128881
360	129600	130321	131044	131769	132496	133225	133956	134689	135424	136161
370	136900	137641	138384	139129	139876	140625	141376	142129	142884	143641
380	144400	145161	145924	146689	147456	148225	148996	149769	150544	151321
390	152100	152881	153664	154449	155236	156025	156816	157609	158404	159201
400	160000	160801	161604	162409	163216	164025	164836	165649	166464	167281
410	168100	168921	169744	170569	171396	172225	173056	173889	174724	175561
420	176400	177241	178084	178929	179776	180625	181476	182329	183184	184041
430	184900	185761	186624	187489	188356	189225	190096	190969	191844	192721
440	193600	194481	195364	196249	197136	198025	198916	199809	200704	201601
450	202500	203401	204304	205209	206116	207025	207936	208849	209764	210681
460	211600	212521	213444	214369	215296	216225	217156	218089	219024	219961
470	220900	221841	222784	223729	224676	225625	226576	227529	228484	229441
480	230400	231361	232324	233289	234256	235225	236196	237169	238144	239121
490	240100	241081	242064	243049	244036	245025	246016	247009	248004	249001

Example: $437^2 = 190969$. Squares of four figure numbers may be computed from the

TABLE XV. SQUARES

	0	1	2	3	4	5	6	7	8	9
500	250000	251001	252004	253009	254016	255025	256036	257049	258064	259081
510	260100	261121	262144	263169	264196	265225	266256	267289	268324	269361
520	270400	271441	272484	273529	274576	275625	276676	277729	278784	279841
530	280900	281961	283024	284089	285156	286225	287296	288369	289444	290521
540	291600	292681	293764	294849	295936	297025	298116	299209	300304	301401
550	302500	303601	304704	305809	306916	308025	309136	310249	311364	312481
560	313600	314721	315844	316969	318096	319225	320356	321489	322624	323761
570	324900	326041	327184	328329	329476	330625	331776	332929	334084	335241
580	336400	337561	338724	339889	341056	342225	343396	344569	345744	346921
590	348100	349281	350464	351649	352836	354025	355216	356409	357604	358801
600	360000	361201	362404	363609	364816	366025	367236	368449	369664	370881
610	372100	373321	374544	375769	376996	378225	379456	380689	381924	383161
620	384400	385641	386884	388129	389376	390625	391876	393129	394384	395641
630	396900	398161	399424	400689	401956	403225	404496	405769	407044	408321
640	409600	410881	412164	413449	414736	416025	417316	418609	419904	421201
650	422500	423801	425104	426409	427716	429025	430336	431649	432964	434281
660	435600	436921	438244	439569	440896	442225	443556	444889	446224	447561
670	448900	450241	451584	452929	454276	455625	456976	458329	459684	461041
680	462400	463761	465124	466489	467856	469225	470596	471969	473344	474721
690	476100	477481	478864	480249	481636	483025	484416	485809	487204	488601
700	490000	491401	492804	494209	495616	497025	498436	499849	501264	502681
710	504100	505521	506944	508369	509796	511225	512656	514089	515524	516961
720	518400	519841	521284	522729	524176	525625	527076	528529	529984	531441
730	532900	534361	535824	537289	538756	540225	541696	543169	544644	546121
740	547600	549081	550564	552049	553536	555025	556516	558009	559504	561001
750	562500	564001	565504	567009	568516	570025	571536	573049	574564	576081
760	577600	579121	580644	582169	583696	585225	586756	588289	589824	591361
770	592900	594441	595984	597529	599076	600625	602176	603729	605284	606841
780	608400	609961	611524	613089	614656	616225	617796	619369	620944	622521
790	624100	625681	627264	628849	630436	632025	633616	635209	636804	638401
800	640000	641601	643204	644809	646416	648025	649636	651249	652864	654481
810	656100	657721	659344	660969	662596	664225	665856	667489	669124	670761
820	672400	674041	675684	677329	678976	680625	682276	683929	685584	687241
830	688900	690561	692224	693889	695556	697225	698896	700569	702244	703921
840	705600	707281	708964	710649	712336	714025	715716	717409	719104	720801
850	722500	724201	725904	727609	729316	731025	732736	734449	736164	737881
860	739600	741321	743044	744769	746496	748225	749956	751689	753424	755161
870	756900	758641	760384	762129	763876	765625	767376	769129	770884	772641
880	774400	776161	777924	779689	781456	783225	784996	786769	788544	790321
890	792100	793881	795664	797449	799236	801025	802816	804609	806404	808201
900	810000	811801	813604	815409	817216	819025	820836	822649	824464	826281
910	828100	829921	831744	833569	835396	837225	839056	840889	842724	844561
920	846400	848241	850084	851929	853776	855625	857476	859329	861184	863041
930	864900	866761	868624	870489	872356	874225	876096	877969	879844	881721
940	883600	885481	887364	889249	891136	893025	894916	896809	898704	900601
950	902500	904401	906304	908209	910116	912025	913936	915849	917764	919681
960	921600	923521	925444	927369	929296	931225	933156	935089	937024	938961
970	940900	942841	944784	946729	948676	950625	952576	954529	956484	958441
980	960400	962361	964324	966289	968256	970225	972196	974169	976144	978121
990	980100	982081	984064	986049	988036	990025	992016	994009	996004	998001

formula $(a \pm b)^2 = a^2 \pm 2ab + b^2$.   Example: $437 \cdot 2^2 = 190969 + 174 \cdot 8 + 0 \cdot 04 = 191143 \cdot 84$.

n	$\sqrt{n}$	$\sqrt{10n}$	n	$\sqrt{n}$	$\sqrt{10n}$	n	$\sqrt{n}$	$\sqrt{10n}$	n	$\sqrt{n}$	$\sqrt{10n}$
1·00	1·0000	3·1623	1·50	1·2247	3·8730	2·00	1·4142	4·4721	2·50	1·5811	5·0000
1·01	1·0050	3·1780	1·51	1·2288	3·8859	2·01	1·4177	4·4833	2·51	1·5843	5·0100
1·02	1·0100	3·1937	1·52	1·2329	3·8987	2·02	1·4213	4·4944	2·52	1·5875	5·0200
1·03	1·0149	3·2094	1·53	1·2369	3·9115	2·03	1·4248	4·5056	2·53	1·5906	5·0299
1·04	1·0198	3·2249	1·54	1·2410	3·9243	2·04	1·4283	4·5166	2·54	1·5937	5·0398
1·05	1·0247	3·2404	1·55	1·2450	3·9370	2·05	1·4318	4·5277	2·55	1·5969	5·0498
1·06	1·0296	3·2558	1·56	1·2490	3·9497	2·06	1·4353	4·5387	2·56	1·6000	5·0596
1·07	1·0344	3·2711	1·57	1·2530	3·9623	2·07	1·4387	4·5497	2·57	1·6031	5·0695
1·08	1·0392	3·2863	1·58	1·2570	3·9749	2·08	1·4422	4·5607	2·58	1·6062	5·0794
1·09	1·0440	3·3015	1·59	1·2610	3·9875	2·09	1·4457	4·5717	2·59	1·6093	5·0892
1·10	1·0488	3·3166	1·60	1·2649	4·0000	2·10	1·4491	4·5826	2·60	1·6125	5·0990
1·11	1·0536	3·3317	1·61	1·2689	4·0125	2·11	1·4526	4·5935	2·61	1·6155	5·1088
1·12	1·0583	3·3466	1·62	1·2728	4·0249	2·12	1·4560	4·6043	2·62	1·6186	5·1186
1·13	1·0630	3·3615	1·63	1·2767	4·0373	2·13	1·4595	4·6152	2·63	1·6217	5·1284
1·14	1·0677	3·3764	1·64	1·2806	4·0497	2·14	1·4629	4·6260	2·64	1·6248	5·1381
1·15	1·0724	3·3912	1·65	1·2845	4·0620	2·15	1·4663	4·6368	2·65	1·6279	5·1478
1·16	1·0770	3·4059	1·66	1·2884	4·0743	2·16	1·4697	4·6476	2·66	1·6310	5·1575
1·17	1·0817	3·4205	1·67	1·2923	4·0866	2·17	1·4731	4·6583	2·67	1·6340	5·1672
1·18	1·0863	3·4351	1·68	1·2961	4·0988	2·18	1·4765	4·6690	2·68	1·6371	5·1769
1·19	1·0909	3·4496	1·69	1·3000	4·1110	2·19	1·4799	4·6797	2·69	1·6401	5·1865
1·20	1·0954	3·4641	1·70	1·3038	4·1231	2·20	1·4832	4·6904	2·70	1·6432	5·1962
1·21	1·1000	3·4785	1·71	1·3077	4·1352	2·21	1·4866	4·7011	2·71	1·6462	5·2058
1·22	1·1045	3·4928	1·72	1·3115	4·1473	2·22	1·4900	4·7117	2·72	1·6492	5·2154
1·23	1·1091	3·5071	1·73	1·3153	4·1593	2·23	1·4933	4·7223	2·73	1·6523	5·2249
1·24	1·1136	3·5214	1·74	1·3191	4·1713	2·24	1·4967	4·7329	2·74	1·6553	5·2345
1·25	1·1180	3·5355	1·75	1·3229	4·1833	2·25	1·5000	4·7434	2·75	1·6583	5·2440
1·26	1·1225	3·5496	1·76	1·3266	4·1952	2·26	1·5033	4·7539	2·76	1·6613	5·2536
1·27	1·1269	3·5637	1·77	1·3304	4·2071	2·27	1·5067	4·7645	2·77	1·6643	5·2631
1·28	1·1314	3·5777	1·78	1·3342	4·2190	2·28	1·5100	4·7749	2·78	1·6673	5·2726
1·29	1·1358	3·5917	1·79	1·3379	4·2308	2·29	1·5133	4·7854	2·79	1·6703	5·2820
1·30	1·1402	3·6056	1·80	1·3416	4·2426	2·30	1·5166	4·7958	2·80	1·6733	5·2915
1·31	1·1446	3·6194	1·81	1·3454	4·2544	2·31	1·5199	4·8062	2·81	1·6763	5·3009
1·32	1·1489	3·6332	1·82	1·3491	4·2661	2·32	1·5232	4·8166	2·82	1·6793	5·3104
1·33	1·1533	3·6469	1·83	1·3528	4·2778	2·33	1·5264	4·8270	2·83	1·6823	5·3198
1·34	1·1576	3·6606	1·84	1·3565	4·2895	2·34	1·5297	4·8374	2·84	1·6852	5·3292
1·35	1·1619	3·6742	1·85	1·3601	4·3012	2·35	1·5330	4·8477	2·85	1·6882	5·3385
1·36	1·1662	3·6878	1·86	1·3638	4·3128	2·36	1·5362	4·8580	2·86	1·6912	5·3479
1·37	1·1705	3·7014	1·87	1·3675	4·3243	2·37	1·5395	4·8683	2·87	1·6941	5·3572
1·38	1·1747	3·7148	1·88	1·3711	4·3359	2·38	1·5427	4·8785	2·88	1·6971	5·3666
1·39	1·1790	3·7283	1·89	1·3748	4·3474	2·39	1·5460	4·8888	2·89	1·7000	5·3759
1·40	1·1832	3·7417	1·90	1·3784	4·3589	2·40	1·5492	4·8990	2·90	1·7029	5·3852
1·41	1·1874	3·7550	1·91	1·3820	4·3704	2·41	1·5524	4·9092	2·91	1·7059	5·3944
1·42	1·1916	3·7683	1·92	1·3856	4·3818	2·42	1·5556	4·9193	2·92	1·7088	5·4037
1·43	1·1958	3·7815	1·93	1·3892	4·3932	2·43	1·5588	4·9295	2·93	1·7117	5·4129
1·44	1·2000	3·7947	1·94	1·3928	4·4045	2·44	1·5620	4·9396	2·94	1·7146	5·4222
1·45	1·2042	3·8079	1·95	1·3964	4·4159	2·45	1·5652	4·9497	2·95	1·7176	5·4314
1·46	1·2083	3·8210	1·96	1·4000	4·4272	2·46	1·5684	4·9598	2·96	1·7205	5·4406
1·47	1·2124	3·8341	1·97	1·4036	4·4385	2·47	1·5716	4·9699	2·97	1·7234	5·4498
1·48	1·2166	3·8471	1·98	1·4071	4·4497	2·48	1·5748	4·9800	2·98	1·7263	5·4589
1·49	1·2207	3·8601	1·99	1·4107	4·4609	2·49	1·5780	4·9900	2·99	1·7292	5·4681

TABLE XVI. SQUARE ROOTS 85

n	$\sqrt{n}$	$\sqrt{10n}$	n	$\sqrt{n}$	$\sqrt{10n}$	n	$\sqrt{n}$	$\sqrt{10n}$	n	$\sqrt{n}$	$\sqrt{10n}$
3·00	1·7321	5·4772	3·50	1·8708	5·9161	4·00	2·0000	6·3246	4·50	2·1213	6·7082
3·01	1·7349	5·4863	3·51	1·8735	5·9245	4·01	2·0025	6·3325	4·51	2·1237	6·7157
3·02	1·7378	5·4955	3·52	1·8762	5·9330	4·02	2·0050	6·3403	4·52	2·1260	6·7231
3·03	1·7407	5·5045	3·53	1·8788	5·9414	4·03	2·0075	6·3482	4·53	2·1284	6·7305
3·04	1·7436	5·5136	3·54	1·8815	5·9498	4·04	2·0100	6·3561	4·54	2·1307	6·7380
3·05	1·7464	5·5227	3·55	1·8841	5·9582	4·05	2·0125	6·3640	4·55	2·1331	6·7454
3·06	1·7493	5·5317	3·56	1·8868	5·9666	4·06	2·0149	6·3718	4·56	2·1354	6·7528
3·07	1·7521	5·5408	3·57	1·8894	5·9749	4·07	2·0174	6·3797	4·57	2·1378	6·7602
3·08	1·7550	5·5498	3·58	1·8921	5·9833	4·08	2·0199	6·3875	4·58	2·1401	6·7676
3·09	1·7578	5·5588	3·59	1·8947	5·9917	4·09	2·0224	6·3953	4·59	2·1424	6·7750
3·10	1·7607	5·5678	3·60	1·8974	6·0000	4·10	2·0248	6·4031	4·60	2·1448	6·7823
3·11	1·7635	5·5767	3·61	1·9000	6·0083	4·11	2·0273	6·4109	4·61	2·1471	6·7897
3·12	1·7664	5·5857	3·62	1·9026	6·0166	4·12	2·0298	6·4187	4·62	2·1494	6·7971
3·13	1·7692	5·5946	3·63	1·9053	6·0249	4·13	2·0322	6·4265	4·63	2·1517	6·8044
3·14	1·7720	5·6036	3·64	1·9079	6·0332	4·14	2·0347	6·4343	4·64	2·1541	6·8118
3·15	1·7748	5·6125	3·65	1·9105	6·0415	4·15	2·0372	6·4420	4·65	2·1564	6·8191
3·16	1·7776	5·6214	3·66	1·9131	6·0498	4·16	2·0396	6·4498	4·66	2·1587	6·8264
3·17	1·7804	5·6303	3·67	1·9157	6·0581	4·17	2·0421	6·4576	4·67	2·1610	6·8337
3·18	1·7833	5·6391	3·68	1·9183	6·0663	4·18	2·0445	6·4653	4·68	2·1633	6·8411
3·19	1·7861	5·6480	3·69	1·9209	6·0745	4·19	2·0469	6·4730	4·69	2·1656	6·8484
3·20	1·7889	5·6569	3·70	1·9235	6·0828	4·20	2·0494	6·4807	4·70	2·1679	6·8557
3·21	1·7916	5·6657	3·71	1·9261	6·0910	4·21	2·0518	6·4885	4·71	2·1703	6·8629
3·22	1·7944	5·6745	3·72	1·9287	6·0992	4·22	2·0543	6·4962	4·72	2·1726	6·8702
3·23	1·7972	5·6833	3·73	1·9313	6·1074	4·23	2·0567	6·5038	4·73	2·1749	6·8775
3·24	1·8000	5·6921	3·74	1·9339	6·1156	4·24	2·0591	6·5115	4·74	2·1772	6·8848
3·25	1·8028	5·7009	3·75	1·9365	6·1237	4·25	2·0616	6·5192	4·75	2·1794	6·8920
3·26	1·8055	5·7096	3·76	1·9391	6·1319	4·26	2·0640	6·5269	4·76	2·1817	6·8993
3·27	1·8083	5·7184	3·77	1·9416	6·1400	4·27	2·0664	6·5345	4·77	2·1840	6·9065
3·28	1·8111	5·7271	3·78	1·9442	6·1482	4·28	2·0688	6·5422	4·78	2·1863	6·9138
3·29	1·8138	5·7359	3·79	1·9468	6·1563	4·29	2·0712	6·5498	4·79	2·1886	6·9210
3·30	1·8166	5·7446	3·80	1·9494	6·1644	4·30	2·0736	6·5574	4·80	2·1909	6·9282
3·31	1·8193	5·7533	3·81	1·9519	6·1725	4·31	2·0761	6·5651	4·81	2·1932	6·9354
3·32	1·8221	5·7619	3·82	1·9545	6·1806	4·32	2·0785	6·5727	4·82	2·1954	6·9426
3·33	1·8248	5·7706	3·83	1·9570	6·1887	4·33	2·0809	6·5803	4·83	2·1977	6·9498
3·34	1·8276	5·7793	3·84	1·9596	6·1968	4·34	2·0833	6·5879	4·84	2·2000	6·9570
3·35	1·8303	5·7879	3·85	1·9621	6·2048	4·35	2·0857	6·5955	4·85	2·2023	6·9642
3·36	1·8330	5·7966	3·86	1·9647	6·2129	4·36	2·0881	6·6030	4·86	2·2045	6·9714
3·37	1·8358	5·8052	3·87	1·9672	6·2209	4·37	2·0905	6·6106	4·87	2·2068	6·9785
3·38	1·8385	5·8138	3·88	1·9698	6·2290	4·38	2·0928	6·6182	4·88	2·2091	6·9857
3·39	1·8412	5·8224	3·89	1·9723	6·2370	4·39	2·0952	6·6257	4·89	2·2113	6·9929
3·40	1·8439	5·8310	3·90	1·9748	6·2450	4·40	2·0976	6·6332	4·90	2·2136	7·0000
3·41	1·8466	5·8395	3·91	1·9774	6·2530	4·41	2·1000	6·6408	4·91	2·2159	7·0071
3·42	1·8493	5·8481	3·92	1·9799	6·2610	4·42	2·1024	6·6483	4·92	2·2181	7·0143
3·43	1·8520	5·8566	3·93	1·9824	6·2690	4·43	2·1048	6·6558	4·93	2·2204	7·0214
3·44	1·8547	5·8652	3·94	1·9849	6·2769	4·44	2·1071	6·6633	4·94	2·2226	7·0285
3·45	1·8574	5·8737	3·95	1·9875	6·2849	4·45	2·1095	6·6708	4·95	2·2249	7·0356
3·46	1·8601	5·8822	3·96	1·9900	6·2929	4·46	2·1119	6·6783	4·96	2·2271	7·0427
3·47	1·8628	5·8907	3·97	1·9925	6·3008	4·47	2·1142	6·6858	4·97	2·2293	7·0498
3·48	1·8655	5·8992	3·98	1·9950	6·3087	4·48	2·1166	6·6933	4·98	2·2316	7·0569
3·49	1·8682	5·9076	3·99	1·9975	6·3166	4·49	2·1190	6·7007	4·99	2·2338	7·0640

n	$\sqrt{n}$	$\sqrt{10n}$	n	$\sqrt{n}$	$\sqrt{10n}$	n	$\sqrt{n}$	$\sqrt{10n}$	n	$\sqrt{n}$	$\sqrt{10n}$	n	$\sqrt{n}$	$\sqrt{10n}$
5·00	2·2361	7·0711	5·50	2·3452	7·4162	6·00	2·4495	7·7460	6·50	2·5495	8·0623	7·00	2·6458	8·3666
5·01	2·2383	7·0781	5·51	2·3473	7·4229	6·01	2·4515	7·7524	6·51	2·5515	8·0685	7·01	2·6476	8·3726
5·02	2·2405	7·0852	5·52	2·3495	7·4297	6·02	2·4536	7·7589	6·52	2·5534	8·0747	7·02	2·6495	8·3785
5·03	2·2428	7·0922	5·53	2·3516	7·4364	6·03	2·4556	7·7653	6·53	2·5554	8·0808	7·03	2·6514	8·3845
5·04	2·2450	7·0993	5·54	2·3537	7·4431	6·04	2·4576	7·7717	6·54	2·5573	8·0870	7·04	2·6533	8·3905
5·05	2·2472	7·1063	5·55	2·3558	7·4498	6·05	2·4597	7·7782	6·55	2·5593	8·0932	7·05	2·6552	8·3964
5·06	2·2494	7·1134	5·56	2·3580	7·4565	6·06	2·4617	7·7846	6·56	2·5612	8·0994	7·06	2·6571	8·4024
5·07	2·2517	7·1204	5·57	2·3601	7·4632	6·07	2·4637	7·7910	6·57	2·5632	8·1056	7·07	2·6589	8·4083
5·08	2·2539	7·1274	5·58	2·3622	7·4699	6·08	2·4658	7·7974	6·58	2·5652	8·1117	7·08	2·6608	8·4143
5·09	2·2561	7·1344	5·59	2·3643	7·4766	6·09	2·4678	7·8038	6·59	2·5671	8·1179	7·09	2·6627	8·4202
5·10	2·2583	7·1414	5·60	2·3664	7·4833	6·10	2·4698	7·8102	6·60	2·5690	8·1240	7·10	2·6646	8·4261
5·11	2·2605	7·1484	5·61	2·3685	7·4900	6·11	2·4718	7·8166	6·61	2·5710	8·1302	7·11	2·6665	8·4321
5·12	2·2627	7·1554	5·62	2·3707	7·4967	6·12	2·4739	7·8230	6·62	2·5729	8·1363	7·12	2·6683	8·4380
5·13	2·2650	7·1624	5·63	2·3728	7·5033	6·13	2·4759	7·8294	6·63	2·5749	8·1425	7·13	2·6702	8·4439
5·14	2·2672	7·1694	5·64	2·3749	7·5100	6·14	2·4779	7·8358	6·64	2·5768	8·1486	7·14	2·6721	8·4499
5·15	2·2694	7·1764	5·65	2·3770	7·5166	6·15	2·4799	7·8422	6·65	2·5788	8·1548	7·15	2·6739	8·4558
5·16	2·2716	7·1833	5·66	2·3791	7·5233	6·16	2·4819	7·8486	6·66	2·5807	8·1609	7·16	2·6758	8·4617
5·17	2·2738	7·1903	5·67	2·3812	7·5299	6·17	2·4839	7·8549	6·67	2·5826	8·1670	7·17	2·6777	8·4676
5·18	2·2760	7·1972	5·68	2·3833	7·5366	6·18	2·4860	7·8613	6·68	2·5846	8·1731	7·18	2·6796	8·4735
5·19	2·2782	7·2042	5·69	2·3854	7·5432	6·19	2·4880	7·8677	6·69	2·5865	8·1792	7·19	2·6814	8·4794
5·20	2·2804	7·2111	5·70	2·3875	7·5498	6·20	2·4900	7·8740	6·70	2·5884	8·1854	7·20	2·6833	8·4853
5·21	2·2825	7·2180	5·71	2·3896	7·5565	6·21	2·4920	7·8804	6·71	2·5904	8·1915	7·21	2·6851	8·4912
5·22	2·2847	7·2250	5·72	2·3917	7·5631	6·22	2·4940	7·8867	6·72	2·5923	8·1976	7·22	2·6870	8·4971
5·23	2·2869	7·2319	5·73	2·3937	7·5697	6·23	2·4960	7·8930	6·73	2·5942	8·2037	7·23	2·6889	8·5029
5·24	2·2891	7·2388	5·74	2·3958	7·5763	6·24	2·4980	7·8994	6·74	2·5962	8·2098	7·24	2·6907	8·5088
5·25	2·2913	7·2457	5·75	2·3979	7·5829	6·25	2·5000	7·9057	6·75	2·5981	8·2158	7·25	2·6926	8·5147
5·26	2·2935	7·2526	5·76	2·4000	7·5895	6·26	2·5020	7·9120	6·76	2·6000	8·2219	7·26	2·6944	8·5206
5·27	2·2956	7·2595	5·77	2·4021	7·5961	6·27	2·5040	7·9183	6·77	2·6019	8·2280	7·27	2·6963	8·5264
5·28	2·2978	7·2664	5·78	2·4042	7·6026	6·28	2·5060	7·9246	6·78	2·6038	8·2341	7·28	2·6981	8·5323
5·29	2·3000	7·2732	5·79	2·4062	7·6092	6·29	2·5080	7·9310	6·79	2·6058	8·2401	7·29	2·7000	8·5381
5·30	2·3022	7·2801	5·80	2·4083	7·6158	6·30	2·5100	7·9373	6·80	2·6077	8·2462	7·30	2·7019	8·5440
5·31	2·3043	7·2870	5·81	2·4104	7·6223	6·31	2·5120	7·9436	6·81	2·6096	8·2523	7·31	2·7037	8·5499
5·32	2·3065	7·2938	5·82	2·4125	7·6289	6·32	2·5140	7·9498	6·82	2·6115	8·2583	7·32	2·7055	8·5557
5·33	2·3087	7·3007	5·83	2·4145	7·6354	6·33	2·5159	7·9561	6·83	2·6134	8·2644	7·33	2·7074	8·5615
5·34	2·3108	7·3075	5·84	2·4166	7·6420	6·34	2·5179	7·9624	6·84	2·6153	8·2704	7·34	2·7092	8·5674
5·35	2·3130	7·3144	5·85	2·4187	7·6485	6·35	2·5199	7·9687	6·85	2·6173	8·2765	7·35	2·7111	8·5732
5·36	2·3152	7·3212	5·86	2·4207	7·6551	6·36	2·5219	7·9750	6·86	2·6192	8·2825	7·36	2·7129	8·5790
5·37	2·3173	7·3280	5·87	2·4228	7·6616	6·37	2·5239	7·9812	6·87	2·6211	8·2885	7·37	2·7148	8·5849
5·38	2·3195	7·3348	5·88	2·4249	7·6681	6·38	2·5259	7·9875	6·88	2·6230	8·2946	7·38	2·7166	8·5907
5·39	2·3216	7·3417	5·89	2·4269	7·6746	6·39	2·5278	7·9937	6·89	2·6249	8·3006	7·39	2·7185	8·5965
5·40	2·3238	7·3485	5·90	2·4290	7·6811	6·40	2·5298	8·0000	6·90	2·6268	8·3066	7·40	2·7203	8·6023
5·41	2·3259	7·3553	5·91	2·4310	7·6877	6·41	2·5318	8·0062	6·91	2·6287	8·3126	7·41	2·7221	8·6081
5·42	2·3281	7·3621	5·92	2·4331	7·6942	6·42	2·5338	8·0125	6·92	2·6306	8·3187	7·42	2·7240	8·6139
5·43	2·3302	7·3689	5·93	2·4352	7·7006	6·43	2·5357	8·0187	6·93	2·6325	8·3247	7·43	2·7258	8·6197
5·44	2·3324	7·3756	5·94	2·4372	7·7071	6·44	2·5377	8·0250	6·94	2·6344	8·3307	7·44	2·7276	8·6255
5·45	2·3345	7·3824	5·95	2·4393	7·7136	6·45	2·5397	8·0312	6·95	2·6363	8·3367	7·45	2·7295	8·6313
5·46	2·3367	7·3892	5·96	2·4413	7·7201	6·46	2·5417	8·0374	6·96	2·6382	8·3427	7·46	2·7313	8·6371
5·47	2·3388	7·3959	5·97	2·4434	7·7266	6·47	2·5436	8·0436	6·97	2·6401	8·3487	7·47	2·7331	8·6429
5·48	2·3409	7·4027	5·98	2·4454	7·7330	6·48	2·5456	8·0498	6·98	2·6420	8·3546	7·48	2·7350	8·6487
5·49	2·3431	7·4095	5·99	2·4474	7·7395	6·49	2·5475	8·0561	6·99	2·6439	8·3606	7·49	2·7368	8·6545

TABLE XVI. SQUARE ROOTS 87

n	$\sqrt{n}$	$\sqrt{10n}$	n	$\sqrt{n}$	$\sqrt{10n}$	n	$\sqrt{n}$	$\sqrt{10n}$	n	$\sqrt{n}$	$\sqrt{10n}$	n	$\sqrt{n}$	$\sqrt{10n}$
7.50	2.7386	8.6603	8.00	2.8284	8.9443	8.50	2.9155	9.2195	9.00	3.0000	9.4868	9.50	3.0822	9.7468
7.51	2.7404	8.6660	8.01	2.8302	8.9499	8.51	2.9172	9.2250	9.01	3.0017	9.4921	9.51	3.0838	9.7519
7.52	2.7423	8.6718	8.02	2.8320	8.9554	8.52	2.9189	9.2304	9.02	3.0033	9.4974	9.52	3.0854	9.7570
7.53	2.7441	8.6776	8.03	2.8337	8.9610	8.53	2.9206	9.2358	9.03	3.0050	9.5026	9.53	3.0871	9.7622
7.54	2.7459	8.6833	8.04	2.8355	8.9666	8.54	2.9223	9.2412	9.04	3.0067	9.5079	9.54	3.0887	9.7673
7.55	2.7477	8.6891	8.05	2.8373	8.9722	8.55	2.9240	9.2466	9.05	3.0083	9.5131	9.55	3.0903	9.7724
7.56	2.7495	8.6948	8.06	2.8390	8.9778	8.56	2.9257	9.2520	9.06	3.0100	9.5184	9.56	3.0919	9.7775
7.57	2.7514	8.7006	8.07	2.8408	8.9833	8.57	2.9275	9.2574	9.07	3.0116	9.5237	9.57	3.0935	9.7826
7.58	2.7532	8.7063	8.08	2.8425	8.9889	8.58	2.9292	9.2628	9.08	3.0133	9.5289	9.58	3.0952	9.7877
7.59	2.7550	8.7121	8.09	2.8443	8.9944	8.59	2.9309	9.2682	9.09	3.0150	9.5341	9.59	3.0968	9.7929
7.60	2.7568	8.7178	8.10	2.8460	9.0000	8.60	2.9326	9.2736	9.10	3.0166	9.5394	9.60	3.0984	9.7980
7.61	2.7586	8.7235	8.11	2.8478	9.0056	8.61	2.9343	9.2790	9.11	3.0183	9.5446	9.61	3.1000	9.8031
7.62	2.7604	8.7293	8.12	2.8496	9.0111	8.62	2.9360	9.2844	9.12	3.0199	9.5499	9.62	3.1016	9.8082
7.63	2.7622	8.7350	8.13	2.8513	9.0167	8.63	2.9377	9.2898	9.13	3.0216	9.5551	9.63	3.1032	9.8133
7.64	2.7641	8.7407	8.14	2.8531	9.0222	8.64	2.9394	9.2952	9.14	3.0232	9.5603	9.64	3.1048	9.8184
7.65	2.7659	8.7464	8.15	2.8548	9.0277	8.65	2.9411	9.3005	9.15	3.0249	9.5656	9.65	3.1064	9.8234
7.66	2.7677	8.7521	8.16	2.8566	9.0333	8.66	2.9428	9.3059	9.16	3.0265	9.5708	9.66	3.1081	9.8285
7.67	2.7695	8.7579	8.17	2.8583	9.0388	8.67	2.9445	9.3113	9.17	3.0282	9.5760	9.67	3.1097	9.8336
7.68	2.7713	8.7636	8.18	2.8601	9.0443	8.68	2.9462	9.3167	9.18	3.0299	9.5812	9.68	3.1113	9.8387
7.69	2.7731	8.7693	8.19	2.8618	9.0499	8.69	2.9479	9.3220	9.19	3.0315	9.5864	9.69	3.1129	9.8438
7.70	2.7749	8.7750	8.20	2.8636	9.0554	8.70	2.9496	9.3274	9.20	3.0332	9.5917	9.70	3.1145	9.8489
7.71	2.7767	8.7807	8.21	2.8653	9.0609	8.71	2.9513	9.3327	9.21	3.0348	9.5969	9.71	3.1161	9.8539
7.72	2.7785	8.7864	8.22	2.8671	9.0664	8.72	2.9530	9.3381	9.22	3.0364	9.6021	9.72	3.1177	9.8590
7.73	2.7803	8.7920	8.23	2.8688	9.0719	8.73	2.9547	9.3434	9.23	3.0381	9.6073	9.73	3.1193	9.8641
7.74	2.7821	8.7977	8.24	2.8705	9.0774	8.74	2.9563	9.3488	9.24	3.0397	9.6125	9.74	3.1209	9.8691
7.75	2.7839	8.8034	8.25	2.8723	9.0830	8.75	2.9580	9.3541	9.25	3.0414	9.6177	9.75	3.1225	9.8742
7.76	2.7857	8.8091	8.26	2.8740	9.0885	8.76	2.9597	9.3595	9.26	3.0430	9.6229	9.76	3.1241	9.8793
7.77	2.7875	8.8148	8.27	2.8758	9.0940	8.77	2.9614	9.3648	9.27	3.0447	9.6281	9.77	3.1257	9.8843
7.78	2.7893	8.8204	8.28	2.8775	9.0995	8.78	2.9631	9.3702	9.28	3.0463	9.6333	9.78	3.1273	9.8894
7.79	2.7911	8.8261	8.29	2.8792	9.1049	8.79	2.9648	9.3755	9.29	3.0480	9.6385	9.79	3.1289	9.8944
7.80	2.7928	8.8318	8.30	2.8810	9.1104	8.80	2.9665	9.3808	9.30	3.0496	9.6437	9.80	3.1305	9.8995
7.81	2.7946	8.8374	8.31	2.8827	9.1159	8.81	2.9682	9.3862	9.31	3.0512	9.6488	9.81	3.1321	9.9045
7.82	2.7964	8.8431	8.32	2.8844	9.1214	8.82	2.9698	9.3915	9.32	3.0529	9.6540	9.82	3.1337	9.9096
7.83	2.7982	8.8487	8.33	2.8862	9.1269	8.83	2.9715	9.3968	9.33	3.0545	9.6592	9.83	3.1353	9.9146
7.84	2.8000	8.8544	8.34	2.8879	9.1324	8.84	2.9732	9.4021	9.34	3.0561	9.6644	9.84	3.1369	9.9197
7.85	2.8018	8.8600	8.35	2.8896	9.1378	8.85	2.9749	9.4074	9.35	3.0578	9.6695	9.85	3.1385	9.9247
7.86	2.8036	8.8657	8.36	2.8914	9.1433	8.86	2.9766	9.4128	9.36	3.0594	9.6747	9.86	3.1401	9.9298
7.87	2.8054	8.8713	8.37	2.8931	9.1488	8.87	2.9783	9.4181	9.37	3.0610	9.6799	9.87	3.1417	9.9348
7.88	2.8071	8.8769	8.38	2.8948	9.1542	8.88	2.9799	9.4234	9.38	3.0627	9.6850	9.88	3.1432	9.9398
7.89	2.8089	8.8826	8.39	2.8965	9.1597	8.89	2.9816	9.4287	9.39	3.0643	9.6902	9.89	3.1448	9.9448
7.90	2.8107	8.8882	8.40	2.8983	9.1652	8.90	2.9833	9.4340	9.40	3.0659	9.6954	9.90	3.1464	9.9499
7.91	2.8125	8.8938	8.41	2.9000	9.1706	8.91	2.9850	9.4393	9.41	3.0676	9.7005	9.91	3.1480	9.9549
7.92	2.8142	8.8994	8.42	2.9017	9.1761	8.92	2.9866	9.4446	9.42	3.0692	9.7057	9.92	3.1496	9.9599
7.93	2.8160	8.9051	8.43	2.9034	9.1815	8.93	2.9883	9.4499	9.43	3.0708	9.7108	9.93	3.1512	9.9649
7.94	2.8178	8.9107	8.44	2.9052	9.1869	8.94	2.9900	9.4552	9.44	3.0725	9.7160	9.94	3.1528	9.9700
7.95	2.8196	8.9163	8.45	2.9069	9.1924	8.95	2.9917	9.4604	9.45	3.0741	9.7211	9.95	3.1544	9.9750
7.96	2.8213	8.9219	8.46	2.9086	9.1978	8.96	2.9933	9.4657	9.46	3.0757	9.7263	9.96	3.1559	9.9800
7.97	2.8231	8.9275	8.47	2.9103	9.2033	8.97	2.9950	9.4710	9.47	3.0773	9.7314	9.97	3.1575	9.9850
7.98	2.8249	8.9331	8.48	2.9120	9.2087	8.98	2.9967	9.4763	9.48	3.0790	9.7365	9.98	3.1591	9.9900
7.99	2.8267	8.9387	8.49	2.9138	9.2141	8.99	2.9983	9.4816	9.49	3.0806	9.7417	9.99	3.1607	9.9950

TABLE XVII. RECIPROCALS

	·00	·01	·02	·03	·04	·05	·06	·07	·08	·09
1·0	1·00000	·99010	·98039	·97087	·96154	·95238	·94340	·93458	·92593	·91743
1·1	·90909	·90090	·89286	·88496	·87719	·86957	·86207	·85470	·84746	·84034
1·2	·83333	·82645	·81967	·81301	·80645	·80000	·79365	·78740	·78125	·77519
1·3	·76923	·76336	·75758	·75188	·74627	·74074	·73529	·72993	·72464	·71942
1·4	·71429	·70922	·70423	·69930	·69444	·68966	·68493	·68027	·67568	·67114
1·5	·66667	·66225	·65789	·65359	·64935	·64516	·64103	·63694	·63291	·62893
1·6	·62500	·62112	·61728	·61350	·60976	·60606	·60241	·59880	·59524	·59172
1·7	·58824	·58480	·58140	·57803	·57471	·57143	·56818	·56497	·56180	·55866
1·8	·55556	·55249	·54945	·54645	·54348	·54054	·53763	·53476	·53191	·52910
1·9	·52632	·52356	·52083	·51813	·51546	·51282	·51020	·50761	·50505	·50251
2·0	·50000	·49751	·49505	·49261	·49020	·48780	·48544	·48309	·48077	·47847
2·1	·47619	·47393	·47170	·46948	·46729	·46512	·46296	·46083	·45872	·45662
2·2	·45455	·45249	·45045	·44843	·44643	·44444	·44248	·44053	·43860	·43668
2·3	·43478	·43290	·43103	·42918	·42735	·42553	·42373	·42194	·42017	·41841
2·4	·41667	·41494	·41322	·41152	·40984	·40816	·40650	·40486	·40323	·40161
2·5	·40000	·39841	·39683	·39526	·39370	·39216	·39063	·38911	·38760	·38610
2·6	·38462	·38314	·38168	·38023	·37879	·37736	·37594	·37453	·37313	·37175
2·7	·37037	·36900	·36765	·36630	·36496	·36364	·36232	·36101	·35971	·35842
2·8	·35714	·35587	·35461	·35336	·35211	·35088	·34965	·34843	·34722	·34602
2·9	·34483	·34364	·34247	·34130	·34014	·33898	·33784	·33670	·33557	·33445
3·0	·33333	·33223	·33113	·33003	·32895	·32787	·32680	·32573	·32468	·32362
3·1	·32258	·32154	·32051	·31949	·31847	·31746	·31646	·31546	·31447	·31348
3·2	·31250	·31153	·31056	·30960	·30864	·30769	·30675	·30581	·30488	·30395
3·3	·30303	·30211	·30120	·30030	·29940	·29851	·29762	·29674	·29586	·29499
3·4	·29412	·29326	·29240	·29155	·29070	·28986	·28902	·28818	·28736	·28653
3·5	·28571	·28490	·28409	·28329	·28249	·28169	·28090	·28011	·27933	·27855
3·6	·27778	·27701	·27624	·27548	·27473	·27397	·27322	·27248	·27174	·27100
3·7	·27027	·26954	·26882	·26810	·26738	·26667	·26596	·26525	·26455	·26385
3·8	·26316	·26247	·26178	·26110	·26042	·25974	·25907	·25840	·25773	·25707
3·9	·25641	·25575	·25510	·25445	·25381	·25316	·25253	·25189	·25126	·25063
4·0	·25000	·24938	·24876	·24814	·24752	·24691	·24631	·24570	·24510	·24450
4·1	·24390	·24331	·24272	·24213	·24155	·24096	·24038	·23981	·23923	·23866
4·2	·23810	·23753	·23697	·23641	·23585	·23529	·23474	·23419	·23364	·23310
4·3	·23256	·23202	·23148	·23095	·23041	·22989	·22936	·22883	·22831	·22779
4·4	·22727	·22676	·22624	·22573	·22523	·22472	·22422	·22371	·22321	·22272
4·5	·22222	·22173	·22124	·22075	·22026	·21978	·21930	·21882	·21834	·21786
4·6	·21739	·21692	·21645	·21598	·21552	·21505	·21459	·21413	·21368	·21322
4·7	·21277	·21231	·21186	·21142	·21097	·21053	·21008	·20964	·20921	·20877
4·8	·20833	·20790	·20747	·20704	·20661	·20619	·20576	·20534	·20492	·20450
4·9	·20408	·20367	·20325	·20284	·20243	·20202	·20161	·20121	·20080	·20040

TABLE XVII. RECIPROCALS

	·00	·01	·02	·03	·04	·05	·06	·07	·08	·09
5·0	·20000	·19960	·19920	·19881	·19841	·19802	·19763	·19724	·19685	·19646
5·1	·19608	·19569	·19531	·19493	·19455	·19417	·19380	·19342	·19305	·19268
5·2	·19231	·19194	·19157	·19120	·19084	·19048	·19011	·18975	·18939	·18904
5·3	·18868	·18832	·18797	·18762	·18727	·18692	·18657	·18622	·18587	·18553
5·4	·18519	·18484	·18450	·18416	·18382	·18349	·18315	·18282	·18248	·18215
5·5	·18182	·18149	·18116	·18083	·18051	·18018	·17986	·17953	·17921	·17889
5·6	·17857	·17825	·17794	·17762	·17731	·17699	·17668	·17637	·17606	·17575
5·7	·17544	·17513	·17483	·17452	·17422	·17391	·17361	·17331	·17301	·17271
5·8	·17241	·17212	·17182	·17153	·17123	·17094	·17065	·17036	·17007	·16998
5·9	·16949	·16920	·16892	·16863	·16835	·16807	·16779	·16750	·16722	·16694
6·0	·16667	·16639	·16611	·16584	·16556	·16529	·16502	·16474	·16447	·16420
6·1	·16393	·16367	·16340	·16313	·16287	·16260	·16234	·16207	·16181	·16155
6·2	·16129	·16103	·16077	·16051	·16026	·16000	·15974	·15949	·15924	·15898
6·3	·15873	·15848	·15823	·15798	·15773	·15748	·15723	·15699	·15674	·15649
6·4	·15625	·15601	·15576	·15552	·15528	·15504	·15480	·15456	·15432	·15408
6·5	·15385	·15361	·15337	·15314	·15291	·15267	·15244	·15221	·15198	·15175
6·6	·15152	·15129	·15106	·15083	·15060	·15038	·15015	·14993	·14970	·14948
6·7	·14925	·14903	·14881	·14859	·14837	·14815	·14793	·14771	·14749	·14728
6·8	·14706	·14684	·14663	·14641	·14620	·14599	·14577	·14556	·14535	·14514
6·9	·14493	·14472	·14451	·14430	·14409	·14388	·14368	·14347	·14327	·14306
7·0	·14286	·14265	·14245	·14225	·14205	·14184	·14164	·14144	·14124	·14104
7·1	·14085	·14065	·14045	·14025	·14006	·13986	·13966	·13947	·13928	·13908
7·2	·13889	·13870	·13850	·13831	·13812	·13793	·13774	·13755	·13736	·13717
7·3	·13699	·13680	·13661	·13643	·13624	·13605	·13587	·13569	·13550	·13532
7·4	·13514	·13495	·13477	·13459	·13441	·13423	·13405	·13387	·13369	·13351
7·5	·13333	·13316	·13298	·13280	·13263	·13245	·13228	·13210	·13193	·13175
7·6	·13158	·13141	·13123	·13106	·13089	·13072	·13055	·13038	·13021	·13004
7·7	·12987	·12970	·12953	·12937	·12920	·12903	·12887	·12870	·12853	·12837
7·8	·12821	·12804	·12788	·12771	·12755	·12739	·12723	·12706	·12690	·12674
7·9	·12658	·12642	·12626	·12610	·12594	·12579	·12563	·12547	·12531	·12516
8·0	·12500	·12484	·12469	·12453	·12438	·12422	·12407	·12392	·12376	·12361
8·1	·12346	·12330	·12315	·12300	·12285	·12270	·12255	·12240	·12225	·12210
8·2	·12195	·12180	·12165	·12151	·12136	·12121	·12107	·12092	·12077	·12063
8·3	·12048	·12034	·12019	·12005	·11990	·11976	·11962	·11947	·11933	·11919
8·4	·11905	·11891	·11876	·11862	·11848	·11834	·11820	·11806	·11792	·11779
8·5	·11765	·11751	·11737	·11723	·11710	·11696	·11682	·11669	·11655	·11641
8·6	·11628	·11614	·11601	·11587	·11574	·11561	·11547	·11534	·11521	·11507
8·7	·11494	·11481	·11468	·11455	·11442	·11429	·11416	·11403	·11390	·11377
8·8	·11364	·11351	·11338	·11325	·11312	·11299	·11287	·11274	·11261	·11249
8·9	·11236	·11223	·11211	·11198	·11186	·11173	·11161	·11148	·11136	·11123
9·0	·11111	·11099	·11086	·11074	·11062	·11050	·11038	·11025	·11013	·11001
9·1	·10989	·10977	·10965	·10953	·10941	·10929	·10917	·10905	·10893	·10881
9·2	·10870	·10858	·10846	·10834	·10823	·10811	·10799	·10787	·10776	·10764
9·3	·10753	·10741	·10730	·10718	·10707	·10695	·10684	·10672	·10661	·10650
9·4	·10638	·10627	·10616	·10604	·10593	·10582	·10571	·10560	·10549	·10537
9·5	·10526	·10515	·10504	·10493	·10482	·10471	·10460	·10449	·10438	·10428
9·6	·10417	·10406	·10395	·10384	·10373	·10363	·10352	·10341	·10331	·10320
9·7	·10309	·10299	·10288	·10277	·10267	·10256	·10246	·10235	·10225	·10215
9·8	·10204	·10194	·10183	·10173	·10163	·10152	·10142	·10132	·10121	·10111
9·9	·10101	·10091	·10081	·10070	·10060	·10050	·10040	·10030	·10020	·10010

TABLE XVIII. LOGARITHMS

	0	1	2	3	4	5	6	7	8	9	1	2	3	4	5	6	7	8
100	0000	0004	0009	0013	0017	0022	0026	0030	0035	0039	0	1	1	2	2	3	3	3
101	0043	0047	0051	0056	0060	0064	0069	0073	0077	0081	1	1	2	2	3	3	3	4
102	0086	0090	0095	0099	0103	0107	0112	0116	0120	0124	0	1	1	2	2	2	3	3
103	0128	0133	0137	0141	0145	0149	0154	0158	0162	0166	0	1	1	2	2	3	3	3
104	0170	0174	0178	0182	0187	0191	0195	0199	0203	0207	1	1	2	2	3	3	3	4
105	0212	0216	0220	0224	0228	0233	0237	0241	0245	0249	0	1	1	2	2	2	3	3
106	0253	0257	0261	0265	0269	0273	0277	0281	0285	0289	1	1	2	2	2	3	3	4
107	0294	0298	0302	0306	0310	0314	0318	0322	0326	0330	0	1	1	2	2	2	3	3
108	0334	0338	0342	0346	0350	0354	0358	0362	0366	0370	1	1	1	2	2	3	3	3
109	0374	0378	0382	0386	0390	0394	0398	0402	0406	0410	1	1	1	2	2	3	3	3
11	0415	0454	0493	0531	0569	0607	0645	0682	0719	0755	3	7	11	15	19	22	26	30
12	0792	0828	0864	0899	0934	0969	1003	1037	1071	1105	4	7	11	14	18	21	25	28
13	1140	1173	1206	1239	1271	1303	1335	1367	1398	1430	3	6	10	13	16	19	23	26
14	1462	1493	1523	1553	1584	1614	1643	1673	1702	1731	3	6	9	12	15	18	21	24
15	1761	1790	1818	1847	1875	1903	1931	1958	1986	2013	3	6	9	12	14	17	20	23
16	2042	2069	2096	2122	2149	2175	2201	2227	2253	2279	2	5	8	10	13	16	18	21
17	2304	2330	2355	2380	2405	2430	2455	2479	2504	2528	3	5	8	10	13	15	18	20
18	2553	2577	2601	2625	2648	2672	2695	2718	2741	2764	2	5	7	9	12	14	16	19
19	2787	2810	2833	2855	2878	2900	2922	2944	2966	2988	3	5	7	9	12	14	16	18
20	3010	3032	3053	3075	3096	3117	3138	3159	3180	3201	3	5	7	9	11	13	15	17
21	3222	3243	3264	3284	3304	3324	3345	3365	3385	3404	2	4	6	8	10	12	14	16
22	3424	3444	3463	3483	3502	3521	3541	3560	3579	3598	2	4	6	8	10	12	14	16
23	3617	3636	3655	3673	3692	3710	3729	3747	3765	3784	2	4	6	8	10	11	13	15
24	3802	3820	3838	3856	3874	3892	3909	3927	3944	3962	2	4	5	7	9	11	13	14
25	3979	3997	4014	4031	4048	4065	4082	4099	4116	4133	2	4	5	7	9	10	12	14
26	4150	4166	4183	4199	4216	4232	4249	4265	4281	4297	2	3	5	7	8	10	12	13
27	4314	4330	4346	4362	4378	4394	4409	4425	4441	4456	1	3	4	6	8	9	11	12
28	4472	4488	4503	4518	4534	4549	4564	4579	4594	4609	1	3	4	6	7	9	10	12
29	4624	4639	4654	4669	4684	4698	4713	4728	4742	4757	1	3	4	6	7	9	10	12
30	4771	4785	4800	4814	4828	4843	4857	4871	4885	4899	2	3	5	6	8	9	10	12
31	4914	4928	4942	4956	4970	4983	4997	5011	5025	5038	1	2	4	5	7	8	9	11
32	5052	5065	5079	5092	5105	5119	5132	5145	5159	5172	1	3	4	5	7	8	9	11
33	5185	5198	5211	5224	5237	5250	5263	5276	5289	5302	2	3	4	5	7	8	9	11
34	5315	5328	5340	5353	5366	5378	5391	5403	5416	5428	1	2	4	5	6	7	9	10
35	5441	5453	5466	5478	5490	5502	5515	5527	5539	5551	1	2	3	5	6	7	8	10
36	5563	5575	5587	5599	5611	5623	5635	5647	5658	5670	1	2	4	5	6	7	8	10
37	5682	5694	5706	5717	5729	5741	5752	5764	5775	5787	1	2	3	4	5	7	8	9
38	5798	5809	5821	5832	5843	5855	5866	5877	5888	5899	1	2	3	5	6	7	8	9
39	5911	5922	5933	5944	5955	5966	5977	5988	5999	6010	1	2	3	4	5	6	8	9
40	6021	6032	6043	6053	6064	6075	6086	6096	6107	6118	1	2	3	4	5	6	7	8
41	6128	6139	6149	6160	6170	6181	6191	6202	6212	6222	1	2	3	4	5	6	7	8
42	6233	6243	6253	6264	6274	6284	6294	6304	6315	6325	1	2	3	4	5	6	7	8
43	6335	6345	6355	6365	6375	6385	6395	6405	6415	6425	1	2	3	4	5	6	7	8
44	6434	6444	6454	6464	6474	6483	6493	6503	6513	6522	1	2	3	4	5	6	7	8
45	6532	6542	6551	6561	6570	6580	6589	6599	6608	6618	1	2	3	4	5	6	7	8
46	6627	6637	6646	6656	6665	6674	6684	6693	6702	6711	1	2	3	4	5	6	7	8
47	6721	6730	6739	6748	6757	6767	6776	6785	6794	6803	1	2	3	4	5	6	7	8
48	6812	6821	6830	6839	6848	6857	6866	6875	6884	6893	1	2	3	4	5	6	7	8
49	6902	6911	6920	6929	6937	6946	6955	6964	6972	6981	1	2	3	3	4	5	6	7

$$\log_e x = 2 \cdot 3026 \log_{10} x.$$

Reproduced from A. K. ERLANG: *Fircifrede Logaritmetavl.*

TABLE XVIII. LOGARITHMS 91

	0	1	2	3	4	5	6	7	8	9		1	2	3	4	5	6	7	8	9
50	6990	6998	7007	7016	7024	7033	7041	7050	7058	7067		1	2	3	4	4	5	6	7	8
51	7075	7084	7092	7101	7109	7118	7126	7135	7143	7151		1	2	3	4	5	5	6	7	8
52	7160	7168	7176	7185	7193	7201	7210	7218	7226	7234		1	2	3	4	4	5	6	7	8
53	7243	7251	7259	7268	7276	7284	7292	7300	7308	7316		1	1	2	3	4	5	5	6	7
54	7324	7332	7340	7348	7356	7364	7372	7380	7388	7396		1	2	2	3	4	5	5	6	7
55	7403	7411	7419	7427	7435	7443	7450	7458	7466	7474		1	2	3	3	4	5	6	7	7
56	7482	7490	7497	7505	7513	7520	7528	7536	7543	7551		1	2	2	3	4	5	5	6	7
57	7559	7566	7574	7582	7589	7597	7604	7612	7619	7627		1	1	2	3	4	4	5	6	7
58	7634	7642	7649	7657	7664	7671	7679	7686	7694	7701		1	2	2	3	4	5	5	6	7
59	7708	7716	7723	7730	7738	7745	7752	7760	7767	7774		1	2	2	3	4	5	5	6	7
60	7782	7789	7796	7803	7810	7818	7825	7832	7839	7846		1	1	2	3	4	4	5	6	6
61	7854	7861	7868	7875	7882	7889	7896	7903	7910	7917		0	1	2	2	3	4	5	5	6
62	7924	7931	7938	7945	7952	7959	7966	7973	7980	7987		0	1	2	3	3	4	5	5	6
63	7993	8000	8007	8014	8021	8028	8034	8041	8048	8055		1	2	2	3	4	4	5	6	6
64	8062	8069	8076	8082	8089	8096	8103	8109	8116	8123		0	1	2	3	4	4	5	6	6
65	8129	8136	8142	8149	8156	8162	8169	8176	8182	8189		1	1	2	3	3	4	5	5	6
66	8195	8202	8208	8215	8221	8228	8234	8241	8247	8254		1	2	2	3	4	4	5	6	6
67	8261	8267	8274	8280	8287	8293	8300	8306	8312	8319		1	1	2	2	3	4	4	5	6
68	8325	8331	8338	8344	8350	8357	8363	8369	8376	8382		1	1	2	3	3	4	5	5	6
69	8389	8395	8401	8408	8414	8420	8426	8433	8439	8445		0	1	2	2	3	3	4	5	5
70	8451	8457	8463	8469	8476	8482	8488	8494	8500	8506		1	1	2	3	3	4	4	5	6
71	8513	8519	8525	8531	8537	8543	8549	8555	8561	8567		1	1	2	2	3	4	4	5	5
72	8573	8579	8585	8591	8597	8603	8609	8615	8621	8627		1	2	2	3	3	4	4	5	6
73	8633	8639	8645	8651	8657	8663	8669	8675	8681	8686		1	1	2	2	3	4	4	5	5
74	8692	8698	8704	8710	8716	8721	8727	8733	8739	8745		1	1	2	2	3	4	4	5	5
75	8751	8757	8763	8768	8774	8780	8786	8791	8797	8803		0	1	1	2	2	3	4	4	5
76	8808	8814	8819	8825	8831	8836	8842	8848	8853	8859		1	1	2	3	3	4	4	5	5
77	8865	8871	8876	8882	8888	8893	8899	8904	8910	8916		0	1	1	2	3	3	4	4	5
78	8921	8927	8932	8938	8943	8949	8954	8960	8965	8971		0	1	2	2	3	3	4	4	5
79	8976	8982	8987	8993	8998	9004	9009	9015	9020	9025		1	1	2	2	3	3	4	4	5
80	9031	9036	9042	9047	9052	9058	9063	9069	9074	9079		1	1	2	2	3	3	4	4	5
81	9085	9090	9095	9101	9106	9111	9117	9122	9127	9133		1	1	2	2	3	3	4	4	5
82	9138	9143	9149	9154	9159	9164	9170	9175	9180	9185		1	1	2	2	3	3	4	4	5
83	9191	9196	9201	9206	9212	9217	9222	9227	9232	9237		1	1	2	2	3	3	4	4	5
84	9243	9248	9253	9259	9264	9269	9274	9279	9284	9289		0	1	1	2	2	3	3	4	4
85	9294	9300	9305	9310	9315	9320	9325	9330	9335	9340		0	1	1	2	2	3	3	4	4
86	9345	9350	9355	9360	9365	9370	9375	9380	9385	9390		1	1	2	2	3	3	4	4	5
87	9395	9400	9405	9410	9415	9420	9425	9430	9435	9440		1	1	2	2	3	3	4	4	5
88	9445	9450	9455	9460	9465	9470	9475	9479	9484	9489		0	1	1	2	2	3	3	4	4
89	9494	9499	9504	9509	9514	9518	9523	9528	9533	9538		0	1	1	2	2	3	3	4	4
90	9543	9547	9552	9557	9562	9567	9571	9576	9581	9586		0	1	1	2	2	3	3	4	4
91	9590	9595	9600	9604	9609	9614	9619	9623	9628	9633		1	1	2	2	3	3	4	4	5
92	9638	9643	9647	9652	9657	9662	9666	9671	9676	9680		0	1	1	2	2	3	3	4	4
93	9685	9690	9694	9699	9704	9708	9713	9717	9722	9727		0	1	1	2	2	3	3	4	4
94	9731	9736	9741	9745	9750	9754	9759	9764	9768	9773		0	1	1	2	2	3	3	4	4
95	9777	9782	9786	9791	9796	9800	9805	9809	9814	9818		0	1	1	2	2	3	3	4	4
96	9823	9827	9832	9836	9841	9845	9850	9854	9859	9863		0	1	1	2	2	3	3	4	4
97	9868	9872	9877	9881	9886	9890	9895	9899	9904	9908		0	1	1	2	2	2	3	3	4
98	9912	9917	9921	9926	9930	9934	9939	9943	9948	9952		0	1	1	2	2	3	3	3	4
99	9956	9960	9965	9969	9974	9978	9982	9987	9991	9995		1	1	2	2	3	3	3	4	4

$$\log_{10} x = 0.43429 \log_e x.$$

G. E. C. Gads Forlag, Copenhagen, by kind permission of the publisher.

TABLE XIX. RANDOM SAMPLING NUMBERS

15	77	01	64	69	69	58	40	81	16	60	20	00	84	22	28	26	46	66	36	86	66	17	34	49
85	40	51	40	10	15	33	94	11	65	57	62	94	04	99	05	57	22	71	77	99	68	12	11	14
47	69	35	90	95	16	17	45	86	29	16	70	48	02	00	59	33	93	28	58	34	32	24	34	07
13	26	87	40	20	40	81	46	08	09	74	99	16	92	99	85	19	01	23	11	74	00	79	41	69
10	55	33	20	47	54	16	86	11	16	59	34	71	55	84	03	48	17	60	13	38	71	23	91	83
05	06	67	26	77	14	85	40	52	68	60	41	94	98	18	62	20	94	03	71	60	26	45	17	92
65	50	89	18	74	42	07	50	15	69	86	97	40	25	88	14	17	73	92	07	93	11	93	45	15
59	68	53	31	55	73	47	16	49	79	69	80	76	16	60	58	53	07	04	53	66	94	94	18	13
31	31	05	36	48	75	16	00	21	11	42	44	84	46	84	83	20	49	17	12	21	93	34	61	16
91	59	46	44	45	49	25	36	12	07	25	90	89	55	25	83	47	17	23	93	99	56	14	39	16
63	59	73	21	67	80	00	25	58	25	72	06	12	86	74	54	79	70	85	88	71	58	21	98	48
89	72	47	46	94	78	56	10	65	97	84	79	42	31	49	94	15	31	13	09	45	43	03	82	81
70	51	21	03	18	50	21	99	49	73	06	99	19	24	96	39	43	10	14	12	94	08	55	54	70
14	15	99	60	44	62	72	38	18	36	63	92	61	55	93	77	66	82	10	91	81	51	67	01	47
92	46	90	39	99	64	08	00	97	27	54	96	63	40	54	34	70	27	48	18	68	59	91	83	32
81	23	17	13	01	37	57	92	16	34	15	80	90	25	64	67	77	29	95	84	80	84	84	87	22
87	54	42	46	56	28	89	02	06	98	59	90	74	13	38	98	66	23	20	23	90	55	31	83	48
74	73	84	98	13	11	48	25	33	39	27	36	08	99	57	60	42	88	68	25	22	89	67	83	16
94	55	14	00	97	32	51	92	47	03	92	33	73	20	21	29	77	37	06	98	64	63	34	31	43
69	21	94	26	20	73	90	70	92	76	49	14	60	34	43	90	51	72	11	07	75	94	19	49	40
82	36	36	89	29	87	70	08	71	98	49	00	89	89	99	29	08	02	72	32	68	16	29	82	19
25	06	22	30	87	87	44	48	90	91	38	53	10	60	29	40	07	58	97	84	09	04	33	56	72
82	37	97	60	92	76	39	17	84	34	67	65	52	89	90	62	97	04	33	81	91	27	56	46	35
83	71	07	22	15	17	55	56	82	62	88	83	86	38	14	63	89	39	81	90	25	62	58	68	87
73	13	79	15	12	18	34	22	24	75	56	47	45	22	81	30	82	38	34	52	57	48	30	34	17
91	28	00	57	30	92	12	38	95	21	15	70	78	50	88	01	07	90	72	77	99	53	04	34	73
33	47	55	62	57	08	21	77	31	05	64	74	04	93	42	20	19	09	71	46	37	32	69	69	89
56	66	25	32	38	64	70	26	27	67	77	40	04	34	63	98	99	89	31	16	12	90	50	28	96
88	40	52	02	29	82	69	34	50	21	74	00	91	27	52	98	72	03	45	65	30	89	71	45	91
87	63	88	23	62	51	07	69	59	02	89	49	14	98	53	41	92	36	07	76	85	37	84	37	47
32	25	21	15	08	82	34	57	57	35	22	03	33	48	84	37	37	29	38	37	89	76	25	09	69
44	61	88	23	13	01	59	47	64	04	99	59	96	20	30	87	31	33	69	45	58	48	00	83	48
94	44	08	67	79	41	61	41	15	60	11	88	83	24	82	24	07	78	61	89	42	58	88	22	16
13	24	40	09	00	65	46	38	61	12	90	62	41	11	59	85	18	42	61	29	88	76	04	21	80
78	27	84	05	99	85	75	67	80	05	57	05	71	70	21	31	99	99	06	96	53	99	25	13	63
42	39	30	02	34	99	46	68	45	15	19	74	15	50	17	44	80	13	86	38	40	45	82	13	44
04	52	43	96	38	13	83	80	72	34	20	84	56	19	49	59	14	85	42	99	71	16	34	33	79
82	85	77	30	16	69	32	46	46	30	84	20	68	72	98	94	62	63	59	44	00	89	06	15	87
38	48	84	88	24	55	46	48	60	06	90	08	83	83	98	40	90	88	25	26	85	74	55	80	85
91	19	05	68	22	58	04	63	21	16	23	38	25	43	32	98	94	65	35	35	16	91	07	12	43
54	81	87	21	31	40	46	17	62	63	99	71	14	12	64	51	68	50	60	78	22	69	51	98	37
65	43	75	12	91	20	36	25	57	92	33	65	95	48	75	00	06	65	25	90	16	29	34	14	43
49	98	71	31	80	59	57	32	43	07	85	06	64	75	27	29	17	06	11	30	68	70	97	87	21
03	98	68	89	39	71	87	32	14	99	42	10	25	37	30	08	27	75	43	97	54	20	69	93	50
56	04	21	34	92	89	81	52	15	12	84	11	12	66	87	47	21	06	86	08	35	39	52	28	09
48	09	36	95	36	20	82	53	32	89	92	68	50	88	17	37	92	02	23	43	63	24	69	80	91
23	97	10	96	57	74	07	95	26	44	93	08	43	30	41	86	45	74	33	78	84	33	38	76	73
43	97	55	45	98	35	69	45	96	80	46	26	39	96	33	60	20	73	30	79	17	19	03	47	28
40	05	08	50	79	89	58	19	86	48	27	98	99	24	08	94	19	15	81	29	82	14	35	88	03
66	97	10	69	02	25	36	43	71	76	00	67	56	12	69	07	89	55	63	31	50	72	20	33	36

TABLE XIX. RANDOM SAMPLING NUMBERS

```
15 62 38 72 92   03 76 09 30 75   77 80 04 24 54   67 60 10 79 26   21 60 03 48 14
77 81 15 14 67   55 24 22 20 55   36 93 67 69 37   72 22 43 46 32   56 15 75 25 12
18 87 05 09 96   45 14 72 41 46   12 67 46 72 02   59 06 17 49 12   73 28 23 52 48
08 58 53 63 66   13 07 04 48 71   39 07 46 96 40   20 86 79 11 81   74 11 15 23 17
16 07 79 57 61   42 19 68 15 12   60 21 59 12 07   04 99 88 22 39   75 16 69 13 84

54 13 05 46 17   05 51 24 53 57   46 51 14 39 17   21 39 89 07 35   47 87 44 36 62
95 27 23 17 39   80 24 44 48 93   75 94 77 09 23   48 75 91 69 03   55 51 09 74 47
22 39 44 74 80   25 95 28 63 90   41 19 48 46 72   51 12 97 39 83   35 83 23 17 29
69 95 21 30 11   98 81 38 00 53   41 40 04 16 78   67 29 83 41 18   30 90 44 37 64
75 75 63 97 12   11 57 05 86 52   82 72 47 72 14   37 72 69 75 48   72 21 52 51 81

08 74 79 30 80   70 11 66 79 25   88 01 94 52 31   38 57 98 71 62   12 56 61 01 54
94 88 45 98 60   90 92 74 77 87   40 18 65 87 37   08 68 62 39 52   84 74 90 68 18
97 35 74 05 75   42 13 49 48 38   74 19 06 42 60   20 79 90 81 77   18 51 71 27 27
53 09 93 28 29   80 19 68 30 45   94 49 49 71 21   93 93 71 30 34   52 65 83 40 13
26 36 68 48 09   37 69 26 22 80   23 34 10 45 70   83 51 07 37 44   62 96 74 42 64

49 16 57 15 79   56 63 22 94 28   11 39 69 55 38   53 06 97 20 42   09 14 90 43 48
03 51 79 78 74   75 23 73 75 98   47 85 07 26 02   61 28 01 22 16   14 12 15 67 22
21 88 87 28 48   23 44 03 03 80   53 89 07 87 93   30 17 84 17 74   16 53 31 39 01
56 41 73 33 41   59 16 59 50 98   24 24 87 06 75   99 52 09 88 05   86 25 43 50 94
72 39 19 70 17   01 04 01 22 33   04 84 63 27 65   84 39 45 55 31   95 88 93 90 37

97 28 25 81 49   71 69 22 04 51   56 46 56 15 10   69 59 99 50 29   33 50 16 93 09
18 87 02 72 08   74 52 16 03 82   20 19 66 23 62   37 51 04 89 31   32 19 59 85 57
53 40 11 75 45   13 56 85 31 37   09 17 71 96 79   39 50 79 27 62   71 14 95 53 03
60 49 03 41 56   78 33 77 28 92   21 90 10 62 01   97 06 45 01 19   95 12 24 18 52
09 16 12 75 04   39 69 95 00 48   26 85 28 73 08   66 92 10 66 75   62 61 27 82 57

64 20 19 87 54   88 15 12 54 24   06 99 57 07 28   51 34 54 98 50   70 88 02 86 48
31 28 07 58 77   03 98 26 76 09   10 44 57 61 28   60 29 85 70 79   80 29 19 98 92
80 04 28 47 76   35 73 67 78 28   09 39 88 63 74   41 26 92 42 33   06 80 06 33 84
24 60 22 51 19   34 54 08 24 73   86 72 11 44 69   76 90 81 17 85   57 47 35 16 84
59 16 11 26 29   18 97 78 44 43   58 92 78 70 80   09 65 32 68 26   65 73 90 50 46

58 54 29 98 27   40 51 92 07 13   58 41 59 56 94   16 32 51 42 54   77 37 13 85 19
20 18 34 22 73   57 40 67 17 28   63 57 74 36 18   65 55 25 50 68   35 90 00 03 38
53 90 46 56 19   50 58 33 84 53   14 74 17 40 73   86 11 04 02 04   02 28 49 62 36
97 16 93 94 65   70 95 95 83 20   91 42 57 95 63   00 86 29 02 53   02 27 86 70 95
72 55 71 70 92   04 22 53 19 29   67 29 13 56 70   45 73 45 05 04   32 43 30 93 41

99 19 72 58 35   49 09 26 00 74   26 42 94 52 02   83 31 85 65 66   31 97 67 52 15
48 21 49 72 97   79 19 64 81 82   78 92 51 96 51   28 79 13 20 82   34 81 39 46 86
52 37 68 15 53   22 98 30 16 31   83 24 87 69 29   24 85 44 25 50   75 62 83 95 41
97 50 52 53 52   26 78 21 68 69   57 79 42 40 89   55 81 75 24 52   51 32 79 97 05
36 05 09 18 11   71 01 63 17 60   11 65 19 43 07   44 86 19 58 92   23 71 32 96 19

20 79 70 09 30   81 14 53 80 93   71 94 10 18 14   83 69 76 53 25   27 36 65 65 05
13 07 89 72 08   00 37 75 14 94   83 85 06 72 66   07 47 30 17 11   16 02 63 97 30
94 26 82 37 43   34 23 00 14 50   96 85 41 17 71   69 20 15 98 82   79 69 68 50 31
13 55 88 38 43   75 37 43 83 85   53 74 54 62 99   68 93 74 43 95   06 26 79 78 87
02 44 24 97 71   97 93 12 70 89   42 52 33 24 91   05 87 53 15 77   49 92 83 97 80

34 90 96 63 54   22 84 36 38 99   85 36 25 03 27   49 24 72 10 50   95 14 18 26 64
13 67 06 34 98   04 20 80 12 54   01 18 54 20 76   92 10 47 04 65   54 45 82 42 90
18 75 55 82 66   34 77 27 71 79   67 65 85 92 68   16 43 83 18 74   12 48 68 87 22
91 25 52 57 15   21 54 40 05 50   67 51 66 45 69   84 72 74 32 30   17 70 40 90 24
76 24 00 14 92   14 29 12 17 73   77 46 44 24 30   48 50 36 30 24   93 08 01 39 37
```

TABLE XIX. RANDOM SAMPLING NUMBERS

```
97 58 55 23 12   87 39 84 32 23   26 91 01 11 26   01 24 06 58 20   33 46 38 86 23
84 95 87 34 95   31 23 12 64 75   89 28 38 15 91   81 89 08 86 08   88 20 02 11 67
11 52 38 09 94   32 47 35 42 67   39 33 89 97 16   28 94 86 93 86   96 13 43 85 99
38 69 94 97 10   44 42 85 46 88   56 56 63 58 22   89 19 26 82 25   94 15 54 65 62
23 99 36 33 41   99 76 22 29 19   92 53 92 15 71   47 57 74 69 03   65 57 90 53 17

09 15 95 74 87   09 63 82 63 29   84 57 45 80 07   13 57 40 58 34   21 93 90 39 21
55 75 91 36 57   38 30 89 64 42   01 84 83 12 79   32 09 56 03 81   90 88 00 71 02
84 62 29 92 42   03 92 37 46 19   90 75 68 84 49   53 80 62 19 20   31 14 42 11 17
79 25 70 07 80   85 32 53 87 11   33 79 14 20 04   12 40 31 74 39   80 21 37 65 20
40 10 91 52 27   21 18 64 61 04   85 55 16 90 71   31 95 15 86 74   87 80 75 71 27

93 18 86 63 72   22 53 44 23 89   38 06 46 04 79   67 77 33 21 75   40 51 74 60 53
63 71 69 30 23   12 85 90 05 07   67 33 56 52 60   21 50 72 26 28   48 67 31 87 61
05 29 95 78 06   10 41 62 18 37   42 91 98 43 33   20 58 62 80 65   19 90 07 84 49
30 04 29 90 89   64 25 66 36 41   99 59 15 43 86   34 10 05 99 83   08 02 18 01 22
75 50 83 42 46   80 76 77 34 16   04 05 06 28 86   60 70 04 13 28   98 76 78 43 69

68 82 44 11 33   11 20 42 00 22   40 03 06 12 45   06 32 34 44 18   01 26 36 78 42
51 38 78 69 65   25 98 73 40 31   12 04 99 51 09   49 04 32 68 68   54 64 15 25 68
98 41 81 63 70   58 43 39 93 18   54 46 98 33 01   47 85 39 81 11   48 84 07 64 76
08 44 37 01 53   59 67 11 11 53   16 98 16 52 52   39 32 22 18 22   04 03 06 77 17
17 30 92 82 09   42 37 88 43 35   11 54 89 05 61   10 46 27 43 33   88 92 72 62 01

74 87 89 10 02   19 45 29 65 70   77 81 98 78 67   05 62 57 08 79   30 32 62 91 87
61 81 52 99 80   11 55 21 98 02   08 26 01 20 16   07 42 88 56 51   31 96 14 85 49
55 08 43 08 22   50 28 03 18 00   80 79 60 18 33   92 36 13 50 41   43 59 82 16 65
44 38 47 15 16   96 03 51 42 15   35 96 40 87 91   56 91 13 58 85   40 06 36 04 30
12 45 97 68 57   62 36 61 03 29   46 60 79 85 99   91 13 99 95 58   75 14 74 88 12

19 95 23 05 45   01 87 81 18 92   36 94 07 14 08   90 32 51 29 61   50 60 34 92 25
71 55 86 72 94   77 08 55 65 50   33 53 94 81 52   36 31 53 12 74   88 59 99 35 95
07 32 94 03 20   66 29 98 75 65   70 30 56 59 08   24 51 75 48 73   11 29 77 08 36
10 35 58 59 25   89 62 60 77 71   24 13 38 20 83   02 48 11 67 95   38 97 15 58 18
62 99 34 08 06   81 46 09 16 82   95 17 13 46 36   51 36 87 56 10   80 79 40 48 82

19 44 35 31 20   16 05 25 26 38   98 94 18 38 88   10 90 29 01 12   48 85 52 97 22
77 76 94 64 49   45 39 58 07 88   32 11 43 09 51   32 69 31 63 02   33 47 08 94 85
97 43 81 59 46   59 26 04 63 86   87 31 55 50 66   11 37 04 68 14   57 17 08 82 48
09 77 93 46 95   36 98 08 77 39   71 44 48 10 19   54 80 24 83 47   06 79 01 78 43
71 09 43 23 16   33 93 21 87 89   16 53 05 53 16   98 96 30 89 49   83 32 23 13 32

25 19 47 70 48   16 91 39 59 80   66 77 96 02 08   59 58 48 91 81   04 31 64 65 15
43 23 23 81 42   61 42 37 17 76   75 40 18 81 33   51 68 04 41 00   72 82 28 68 03
50 57 81 53 79   98 04 75 77 30   49 18 17 01 70   06 01 53 04 76   49 93 39 68 00
81 04 78 50 20   33 21 64 10 00   49 43 08 86 53   25 50 24 70 63   01 08 52 66 67
19 62 59 60 23   26 11 30 12 63   26 60 61 15 83   27 41 02 61 80   72 19 91 56 53

32 52 48 94 61   60 43 08 29 67   86 20 90 03 18   48 22 42 82 59   84 31 00 92 15
79 73 88 64 27   89 92 95 64 78   40 06 16 28 66   54 93 14 19 00   39 11 13 27 55
05 12 93 24 38   18 25 64 65 51   81 15 80 43 36   94 49 89 58 80   80 76 25 65 69
59 72 45 18 64   49 67 78 83 66   72 92 63 42 78   21 14 35 00 16   05 92 74 20 31
22 75 30 52 34   00 43 50 50 91   10 64 18 60 30   48 99 84 23 37   20 03 50 50 05

86 21 48 23 45   01 80 49 33 99   57 92 46 06 55   60 98 81 40 20   72 45 67 83 67
47 02 27 40 96   41 44 06 54 76   83 52 32 56 15   09 45 22 54 07   49 70 54 48 84
36 76 21 72 44   85 55 63 87 29   62 84 18 48 29   23 75 29 90 68   02 56 04 32 34
43 84 04 45 20   18 42 25 25 95   70 15 92 80 82   47 10 21 18 57   83 54 02 09 53
88 82 00 84 16   82 67 66 77 89   78 31 98 11 56   27 07 76 59 71   87 56 99 27 28
```

TABLE XIX. RANDOM SAMPLING NUMBERS

```
90 78 82 54 47   20 83 80 10 41   35 22 23 03 98   79 74 41 35 05   78 73 95 47 83
78 58 68 87 41   11 08 81 29 89   71 23 10 01 79   25 06 00 45 80   64 70 95 34 29
51 42 21 03 88   20 05 35 93 00   68 12 09 55 09   36 54 95 22 82   48 30 09 56 87
93 15 07 60 86   67 37 94 24 35   82 44 19 92 96   21 84 29 04 29   83 32 05 10 48
27 12 31 66 62   09 54 17 31 23   27 30 37 36 79   75 50 39 57 12   67 23 22 09 33

79 44 83 55 47   96 50 93 56 82   58 16 35 18 87   64 08 22 47 93   86 43 43 30 17
89 73 43 91 03   57 91 35 40 64   13 61 94 37 16   09 93 96 25 87   30 23 42 54 31
29 30 90 00 58   15 99 93 33 67   80 08 59 21 66   13 54 56 85 25   05 32 03 52 52
97 33 17 26 25   04 73 18 10 05   34 40 32 65 07   28 68 29 31 97   89 57 95 55 16
07 15 44 92 47   28 50 93 03 53   37 70 19 68 59   95 39 87 90 46   98 64 46 24 71

82 50 35 50 80   23 67 81 25 02   83 08 12 70 00   25 31 33 80 06   19 86 14 59 27
59 21 86 16 30   27 85 16 26 34   50 15 87 22 69   71 36 95 90 76   90 99 79 63 21
04 19 60 33 05   29 02 33 74 56   38 84 21 07 35   93 54 70 18 47   14 62 75 45 02
96 91 44 09 94   06 89 50 88 83   82 50 11 82 51   30 68 91 06 28   86 65 17 45 20
31 71 03 53 38   94 02 52 72 15   44 49 53 42 43   00 36 97 67 64   12 27 46 00 18

03 70 22 67 59   98 10 64 68 08   79 06 89 48 41   85 72 10 87 24   96 04 20 68 00
08 45 79 46 89   74 73 67 60 15   70 37 61 44 07   67 89 81 54 26   57 17 63 27 74
37 80 05 75 64   48 51 68 68 27   71 75 45 32 27   76 35 26 58 88   67 74 48 90 94
90 63 56 69 37   19 74 48 63 31   52 36 84 40 66   72 66 03 41 87   65 29 12 36 64
22 69 38 02 88   89 71 43 01 87   41 79 42 99 29   41 08 47 32 19   45 29 59 69 90

05 79 69 67 64   36 14 82 65 26   40 51 63 42 48   85 48 34 12 04   33 26 52 26 52
48 91 53 03 82   64 24 06 31 03   97 44 82 24 89   88 48 66 54 10   41 27 09 11 61
94 64 97 27 25   62 23 94 40 54   56 32 97 78 90   58 86 41 75 19   42 90 85 36 68
15 85 82 52 08   52 96 26 92 88   93 11 03 23 52   78 23 57 85 43   53 90 42 22 22
09 81 37 66 56   99 08 59 19 48   29 69 21 64 95   12 08 15 24 45   59 25 22 76 96

43 83 99 02 76   12 16 45 52 66   35 70 93 09 52   75 40 34 35 62   65 42 27 20 59
31 98 09 80 62   75 26 64 57 26   46 41 47 90 97   99 46 10 51 42   73 28 98 89 91
81 35 42 62 84   37 02 59 78 16   17 96 05 71 39   88 05 34 05 92   22 43 89 66 89
97 95 56 39 75   65 47 61 86 33   14 88 55 33 69   70 87 79 94 46   17 61 72 27 01
37 63 35 93 23   17 30 14 51 51   17 28 21 74 67   12 11 57 19 27   38 70 73 82 92

39 22 96 00 48   52 49 62 09 40   08 30 27 54 70   96 06 52 12 80   36 12 38 68 05
61 29 84 34 51   60 19 77 82 16   64 45 02 27 04   65 55 90 95 04   20 39 29 96 28
38 84 18 10 29   19 09 66 06 78   37 09 60 50 21   52 72 01 52 70   29 65 05 37 16
64 29 48 04 08   55 72 25 25 77   54 26 27 24 39   66 67 06 40 00   99 35 70 69 58
64 02 32 99 63   62 42 89 32 20   81 14 08 40 45   22 15 37 49 38   96 51 19 08 27

13 83 39 51 30   31 49 94 83 66   02 50 95 18 98   58 84 90 58 81   00 40 91 12 46
83 30 90 09 35   41 12 87 93 66   85 96 20 65 34   23 13 05 41 01   91 48 95 59 45
46 63 53 97 63   18 86 37 56 20   35 62 66 11 37   30 91 89 97 51   64 78 06 95 65
54 43 40 02 41   55 70 52 96 87   02 82 61 21 88   60 65 98 42 09   03 61 20 83 01
27 18 65 62 01   97 45 79 51 37   74 47 20 11 48   97 93 73 86 50   46 61 95 01 24

45 42 16 13 20   34 51 08 71 52   39 17 71 39 84   97 27 72 49 42   81 62 32 87 22
35 92 97 02 34   93 32 95 81 13   92 05 40 70 95   71 66 61 24 08   77 32 73 66 79
60 55 35 57 24   52 95 84 90 64   38 39 72 70 17   98 42 85 96 67   41 11 83 17 78
43 17 21 09 60   58 86 12 31 11   66 61 43 96 00   93 97 00 15 20   37 96 73 56 63
07 85 74 58 28   38 74 68 32 61   87 14 71 83 47   90 11 96 70 08   67 04 34 46 08

33 00 29 08 87   42 59 40 24 97   44 99 13 56 87   95 02 47 97 89   23 51 45 37 83
97 14 00 42 23   72 03 19 02 41   11 23 36 98 32   19 91 42 03 58   62 23 74 45 06
68 58 32 80 82   40 49 71 83 37   93 49 99 60 72   88 14 26 88 95   48 69 35 40 63
39 87 38 16 06   82 92 62 32 75   67 64 50 49 39   29 55 53 92 97   04 48 60 53 90
37 73 01 84 87   42 88 30 93 75   01 18 34 73 30   28 44 28 18 01   00 38 26 38 57
```

TABLE XIX. RANDOM SAMPLING NUMBERS

83	10	03	87	35	31	89	45	64	40	61	57	87	29	73	69	40	81	83	78	38	48	42	22	29
36	57	72	32	39	32	78	83	08	53	87	69	15	29	29	11	24	78	09	01	35	27	12	80	54
63	83	40	96	33	52	74	75	58	02	66	94	42	87	71	15	33	56	74	49	92	05	98	56	05
85	29	24	77	88	45	67	31	74	66	76	46	33	84	55	50	13	03	00	97	36	18	25	57	31
91	49	28	72	30	03	03	32	57	95	19	01	67	59	15	00	72	44	40	07	93	42	69	48	89
87	60	75	26	58	65	28	70	09	88	76	32	66	08	99	67	99	07	26	67	47	67	42	72	31
93	20	96	93	05	95	71	90	83	58	02	47	28	10	24	80	64	66	52	38	34	94	17	54	28
94	34	73	87	35	53	33	66	29	50	21	38	74	75	20	54	87	63	74	34	80	48	64	70	79
10	42	12	47	60	34	51	84	25	61	15	76	23	67	84	81	43	35	46	20	95	21	48	80	57
42	57	65	70	17	49	44	83	03	32	53	60	38	20	38	27	02	61	96	31	36	84	25	44	05
14	80	92	91	76	39	57	07	38	03	12	39	71	36	50	82	44	43	32	84	37	77	36	93	64
03	00	22	34	59	83	22	53	06	87	63	11	38	68	14	69	63	25	22	53	49	99	06	31	53
35	10	31	74	97	55	56	24	84	81	54	14	11	24	60	97	32	52	97	29	86	25	07	61	67
75	65	43	98	76	14	42	68	43	55	85	35	29	44	75	95	08	42	39	30	34	12	53	00	91
42	44	24	72	86	52	95	61	21	88	53	57	56	76	42	39	29	24	47	87	06	89	29	41	09
88	16	39	38	57	70	64	36	88	96	88	80	42	18	56	79	44	06	33	07	20	64	21	48	29
13	30	95	40	57	41	19	03	38	11	74	30	45	56	81	14	84	18	09	31	20	98	29	04	09
23	36	64	70	84	35	90	33	28	80	31	42	17	42	14	53	97	26	51	79	56	39	88	41	10
41	86	43	21	65	27	78	35	17	70	05	54	68	00	81	57	73	00	65	27	37	88	17	32	78
23	77	82	23	63	32	70	14	93	29	92	35	32	80	34	12	77	83	97	42	98	41	16	08	33
65	67	57	25	37	61	14	90	95	89	63	15	38	22	88	67	54	24	16	70	36	03	43	78	99
67	31	48	80	29	29	75	62	66	20	06	95	93	05	22	42	18	68	82	07	35	85	84	11	10
94	07	25	78	88	71	92	56	13	38	00	21	32	91	42	57	87	41	13	39	46	60	33	81	89
98	40	03	05	13	23	79	44	10	06	62	37	35	74	89	37	32	87	50	25	44	94	20	81	69
99	28	14	84	72	37	60	08	57	93	31	46	40	68	65	88	96	64	46	69	09	77	52	93	91
50	07	12	34	37	43	89	16	85	44	26	59	20	40	26	77	28	35	71	03	00	56	60	15	16
43	43	27	67	69	24	70	00	71	43	41	27	40	43	75	26	53	43	14	09	54	03	37	54	29
23	71	33	20	98	75	46	75	11	95	11	27	73	15	47	01	53	39	55	47	27	77	51	39	16
33	39	03	80	68	63	76	38	21	28	89	65	05	32	04	09	18	15	91	48	77	27	49	07	59
70	60	79	38	93	80	47	18	80	30	23	27	51	39	11	56	13	89	58	19	42	45	52	60	02
42	80	22	71	37	41	78	34	80	50	52	35	37	72	51	91	15	29	08	28	59	22	58	29	34
70	15	04	92	81	58	71	82	88	34	67	53	42	67	05	60	96	28	14	96	88	77	62	92	69
89	87	13	76	96	78	80	29	11	43	96	96	75	43	06	24	08	07	41	67	81	13	28	85	66
34	71	21	83	96	26	78	18	33	22	07	54	10	84	39	64	03	35	82	03	68	81	64	19	12
01	56	07	44	40	86	57	76	42	27	84	48	21	76	80	00	59	65	93	84	58	68	72	17	66
17	09	69	35	42	26	01	07	58	52	42	61	78	67	23	91	24	18	12	18	09	62	97	22	36
41	91	77	30	25	56	04	03	64	48	31	09	01	73	10	68	86	74	11	01	49	69	39	17	69
36	07	08	01	86	46	94	22	08	18	17	57	91	42	64	35	92	91	01	66	43	36	00	26	20
05	84	14	25	59	55	02	52	38	43	05	06	95	35	82	87	01	42	16	65	82	20	99	51	49
95	65	59	54	30	45	52	24	91	19	53	64	17	80	74	25	76	50	14	95	68	70	26	16	74
02	86	60	83	38	23	99	89	05	41	87	87	85	39	40	95	60	08	08	85	03	56	91	38	35
38	42	83	19	25	46	11	48	40	62	08	44	36	60	81	86	80	71	73	34	55	45	64	10	13
85	91	80	69	35	56	79	85	78	37	59	19	01	02	02	54	28	98	50	14	87	81	68	59	64
12	14	28	93	12	12	71	38	81	60	79	92	19	78	49	56	14	93	03	40	38	33	83	93	39
17	80	74	77	75	62	70	70	90	72	12	96	21	03	70	90	62	53	23	36	67	00	59	11	68
06	50	91	31	95	98	95	81	72	03	37	09	32	11	61	35	40	44	65	14	15	78	17	94	53
26	74	80	36	23	70	43	60	72	48	13	53	12	03	87	46	80	87	39	20	09	69	58	42	84
31	60	92	27	82	42	36	64	13	90	83	89	57	63	85	98	80	30	06	82	85	82	64	86	77
32	68	19	58	65	03	76	48	17	22	61	61	51	26	36	97	54	95	79	57	59	31	78	34	27
90	84	21	49	55	99	37	42	96	82	94	55	34	23	57	13	95	33	65	37	77	37	43	21	41

TABLE XIX. RANDOM SAMPLING NUMBERS

```
05 97 03 44 96   49 87 13 01 69   56 19 27 30 31   16 95 62 74 55   62 27 80 55 00
78 07 89 48 22   97 68 14 69 88   21 52 73 88 28   08 27 09 48 86   34 88 92 17 36
24 08 18 32 54   17 10 47 62 05   13 45 48 09 65   88 39 09 21 48   12 46 88 81 22
02 62 56 96 61   58 40 50 70 94   07 38 72 82 85   68 21 50 32 70   50 99 54 26 39
65 85 71 69 31   00 03 29 96 17   55 28 81 32 56   58 65 37 17 49   62 61 84 05 49

96 13 18 60 79   62 09 04 94 05   03 75 77 32 37   13 74 76 59 41   60 49 08 17 32
83 73 96 86 33   43 15 91 12 62   49 52 62 12 27   85 64 63 67 13   72 58 85 85 79
37 10 24 50 84   63 32 25 04 92   56 86 34 21 10   46 28 41 11 27   65 01 80 44 11
83 63 98 19 87   48 93 24 60 58   14 20 24 31 90   06 21 43 33 17   01 06 38 29 32
31 28 23 11 97   92 99 45 16 23   88 32 80 13 98   26 36 83 38 31   40 86 82 06 47

01 55 37 09 08   81 17 93 34 68   48 50 08 41 90   60 26 43 21 86   50 58 48 36 73
88 62 21 77 17   19 83 36 01 24   85 14 14 31 77   37 05 31 81 51   18 19 62 50 12
42 94 20 19 46   97 74 57 42 36   41 15 88 09 02   71 08 54 31 04   44 33 74 61 78
02 94 74 35 47   44 49 19 83 14   64 35 71 96 78   39 82 22 51 72   72 82 37 27 37
98 93 56 53 31   43 87 88 26 07   51 33 57 32 67   94 12 59 72 70   40 60 24 24 88

04 69 96 20 56   34 82 13 40 44   35 73 12 04 47   89 03 15 13 38   57 23 90 10 64
46 17 32 80 88   50 94 76 71 00   66 80 63 10 19   43 71 14 39 67   18 68 83 99 86
03 03 53 25 33   48 56 38 29 97   72 26 86 66 59   45 72 22 50 57   85 61 96 34 51
51 93 74 83 62   36 47 66 48 33   77 88 04 09 60   16 52 91 29 36   82 02 81 08 41
12 57 43 34 86   58 28 03 45 04   17 56 69 15 72   92 11 29 73 60   32 39 72 92 29

53 84 34 35 10   80 58 23 33 53   41 31 36 65 19   36 95 29 39 82   36 44 83 29 54
55 38 24 04 52   34 62 02 75 71   31 79 80 57 90   83 04 81 87 06   39 94 97 18 98
44 18 54 27 28   95 03 69 23 00   22 27 93 81 82   27 62 41 14 88   15 55 08 23 49
70 40 80 38 20   65 19 35 17 50   53 98 92 85 83   17 17 27 36 58   28 81 79 50 86
04 73 99 66 15   80 49 90 75 98   81 37 70 45 42   88 08 71 02 61   36 66 22 05 45

35 42 90 60 81   67 17 29 99 11   27 47 34 75 24   58 34 16 99 76   47 96 74 27 60
83 09 18 41 03   44 38 21 48 83   47 02 60 64 06   02 79 41 48 98   70 99 85 86 05
70 73 27 11 03   32 39 41 95 31   48 92 07 52 74   04 60 86 62 54   74 82 96 12 31
52 54 07 60 09   11 28 74 68 44   65 54 62 25 13   90 38 80 51 09   63 66 08 69 63
76 73 21 52 45   47 21 92 11 66   09 43 96 59 57   34 35 27 05 22   59 22 31 16 52

82 51 57 91 88   66 47 82 53 13   27 84 57 65 23   27 64 73 11 55   49 12 71 94 95
95 81 86 44 22   67 86 41 16 79   76 11 42 25 48   69 34 95 67 54   37 70 36 35 96
71 15 85 43 23   14 87 31 12 31   46 97 62 75 56   98 94 05 25 97   77 35 25 77 47
68 51 40 99 65   59 34 38 61 88   74 60 22 09 94   69 36 04 99 84   57 92 33 78 45
55 33 10 60 67   39 62 16 35 78   84 93 94 14 10   56 11 50 64 11   13 78 75 52 94

43 56 50 63 55   60 09 73 81 27   36 83 55 12 43   83 05 88 57 87   96 61 34 01 50
19 60 92 82 67   31 06 48 13 20   59 96 66 97 84   72 73 16 83 42   31 61 35 11 32
00 01 71 82 97   15 55 79 08 53   56 52 49 15 44   16 86 30 72 18   54 09 89 88 36
39 18 47 54 33   55 91 02 90 29   81 74 34 98 16   95 03 65 94 61   87 12 02 44 75
60 74 52 41 39   70 85 23 26 19   13 27 95 69 73   54 97 86 65 16   23 66 41 51 77

08 33 98 06 13   75 58 12 37 63   00 43 89 06 97   92 67 83 52 14   40 24 35 63 04
90 78 83 08 33   13 86 55 41 35   26 91 94 70 72   10 46 68 93 19   63 53 12 78 48
55 77 40 85 45   71 63 09 79 12   53 67 21 77 12   01 16 30 72 76   47 79 27 39 59
32 37 88 40 16   08 22 32 48 18   85 65 10 25 54   18 45 24 66 79   89 75 88 70 59
86 14 97 98 50   80 62 48 58 87   03 56 33 46 00   23 79 76 36 62   77 78 78 29 17

82 01 52 21 90   27 95 27 62 34   99 79 54 37 02   69 99 83 70 68   07 39 06 50 24
99 87 92 23 49   06 96 93 21 97   59 73 66 29 74   39 30 89 05 10   02 57 59 89 34
37 67 24 02 62   68 25 66 66 24   48 12 13 94 93   47 54 64 03 40   66 52 45 98 42
44 24 63 02 09   81 52 96 73 00   20 67 70 19 65   80 69 01 80 47   36 75 39 66 96
60 61 42 61 47   50 35 25 28 26   66 25 62 99 76   45 73 32 96 07   64 46 87 73 66
```